Management and Industrial Engineering

This series fosters information exchange and discussion on management and industrial engineering and related aspects, namely global management, organizational development and change, strategic management, lean production, performance management, production management, quality engineering, maintenance management, productivity improvement, materials management, human resource management, workforce behavior, innovation and change, technological and organizational flexibility, self-directed work teams, knowledge management, organizational learning, learning organizations, entrepreneurship, sustainable management, etc. The series provides discussion and the exchange of information on principles, strategies, models, techniques, methodologies and applications of management and industrial engineering in the field of the different types of organizational activities. It aims to communicate the latest developments and thinking in what concerns the latest research activity relating to new organizational challenges and changes world-wide. Contributions to this book series are welcome on all subjects related with management and industrial engineering. To submit a proposal or request further information, please contact Professor J. Paulo Davim, Book Series Editor, pdavim@ua.pt

Mangey Ram · Om Ji Shukla · Kishore Debnath
Editors

Technological Advancements in Engineering and Management Sciences

Springer

Editors
Mangey Ram
Department of Mathematics
Graphic Era Deemed to be University
Dehradun, Uttarakhand, India

Om Ji Shukla
Department of Mechanical Engineering
National Institute of Technology Patna
Patna, Bihar, India

Kishore Debnath
Department of Mechanical Engineering
National Institute of Technology Meghalaya
Shillong, Meghalaya, India

ISSN 2365-0532 ISSN 2365-0540 (electronic)
Management and Industrial Engineering
ISBN 978-3-032-04605-5 ISBN 978-3-032-04606-2 (eBook)
https://doi.org/10.1007/978-3-032-04606-2

This Springer imprint is published by the registered company Springer Nature Switzerland AG
The registered company address is: Gewerbestrasse 11, 6330 Cham, Switzerland

Preface

In an era defined by rapid technological evolution and unprecedented global challenges, the fusion of engineering and management sciences stands as a critical force shaping industries, societies, and the future of innovation. The book *Technological Advancements in Engineering and Management Sciences* is a curated compendium of contemporary research, practical insights, and visionary perspectives that reflect the dynamic intersections between these two vital domains.

This edited volume brings together contributions from scholars, researchers, and professionals across diverse disciplines and geographies, offering a multifaceted view of how technological progress is reshaping engineering practices and management strategies alike. From cutting-edge developments in automation, artificial intelligence, and sustainable infrastructure, to novel approaches in organizational leadership, supply chain management, and data-driven decision-making, each chapter encapsulates the essence of transformation and progress.

The primary goal of this book is to serve as a platform for knowledge dissemination and intellectual exchange among academics, practitioners, and students. By bridging theoretical advancements with real-world applications, it aspires to foster a deeper understanding of the synergies between technology, engineering systems, and managerial effectiveness in a globally connected world.

We extend our sincere gratitude to all the authors and reviewers who contributed their time, expertise, and insights to make this volume possible. Their work not only enriches the academic discourse but also inspires innovation and forward-thinking across disciplines.

It is our hope that this book will serve as a valuable resource for readers seeking to explore the latest trends, challenges, and opportunities in the ever-evolving fields of engineering and management sciences.

Dehradun, India — Prof. Mangey Ram
Patna, India — Dr. Om Ji Shukla
Shillong, India — Dr. Kishore Debnath

Contents

About the Editors

Prof. Mangey Ram received the Ph.D. degree major in Mathematics and minor in Computer Science from G. B. Pant University of Agriculture and Technology, Pantnagar, India in 2008. He has been the Faculty Member for around 17 years and has taught several core courses in pure and applied mathematics at undergraduate, postgraduate, and doctorate levels. He is currently the *Research Professor* and *Dean* (*Research Collaborations*) at *Graphic Era Deemed to be University, Dehradun, India.* Before joining the Graphic Era University, he was the Deputy Manager (Probationary Officer) with Syndicate Bank for a short period. Professor Ram is Editor-in-Chief of *International Journal of Mathematical, Engineering and Management Sciences*; *Journal of Reliability and Statistical Studies*; *Journal of Graphic Era University*; Series Editor of six Book Series with *Elsevier, CRC Press-A Taylor and Frances Group, Walter De Gruyter Publisher Germany, River Publishers* and the Guest Editor and Associate Editor with various journals. He has published 400 plus publications (journal articles/books/book chapters/conference articles) in IEEE, Taylor & Francis, Springer Nature, Elsevier, Emerald, World Scientific and many other national and international journals and conferences. Also, he has published more than 60 books (authored/edited) with international publishers like Elsevier, Springer Nature, CRC Press-A Taylor and Frances Group, Walter De Gruyter Publisher Germany, River Publisher. His fields of research are reliability theory and applied mathematics. Dr. Ram is a *Senior Member* of the IEEE, *Senior Life Member* of Operational Research Society of India, Society for Reliability Engineering, Quality and Operations Management in India, Indian Society of Industrial and Applied Mathematics, He has been a member of the organizing committee of a number of international and national conferences, seminars, and workshops. He has been conferred with "*Young Scientist Award*" by the Uttarakhand State Council for Science and Technology, Dehradun, in 2009. He has been awarded the "Best Faculty Award" in 2011; "*Research Excellence Award*" in 2015; "*Outstanding Researcher Award*" in 2018 for his significant contribution in academics and research at Graphic Era Deemed to be University, Dehradun, India. Recently, he has been received the "*Excellence in Research of the Year-2021 Award*" by the Honourable Chief Minister of Uttarakhand State, India and "*Emerging Mathematician of Uttarakhand-2022*" award by the Director, Directorate of Higher Education Uttarakhand at Uttarakhand Open University, Haldwani, India.

Dr. Om Ji Shukla presently working as an Assistant Professor in the Department of Mechanical Engineering of National Institute of Technology Patna. He holds a Master's degree from BITS Pilani and Ph.D. from Malaviya National Institute of Technology (MNIT) Jaipur. His research interests are Agent based modelling, Artificial Intelligence applications in manufacturing, Data Analytics, Operations management, Supply Chain Management, and E-waste management. He has guided 15 B.Tech. projects, three M.Tech. Dissertations and two Ph.D. Theses. At present, eight Ph.D. Students are working under his supervision. Currently his research team is working in the area of Food supply chain, E-Waste management, Vaccine supply chain, Millet supply chain etc. He has published more than 40 research papers in International Journals, International conference proceedings and book chapters. He is also working in different roles like Associate Dean (Student Welfare), Regional Coordinator of Unnat Bharat Abhiyaan in Bihar (Flagship Program of Ministry of Education (MoE), GoI), Prof. In-Charge of Entrepreneurship Cell. He has been nominated as a member of the Organizing Committee and an Expert of Students' Innovation Festival-IISF-2022 organized by Ministry of Science and Technology, GoI at MANIT, Bhopal.

Dr. Kishore Debnath presently working as an Associate Professor in the Department of Mechanical Engineering of National Institute of Technology Meghalaya completed his bachelor's and master's degrees in mechanical engineering from National Institute of Technology Agartala and National Institute of Technology Rourkela, respectively. He completed his Ph.D. from IIT Roorkee in 2015 in the field of manufacturing of composite materials. He has published more than 90 manuscripts in peer-reviewed journals. He has also contributed several chapters in different books published by internationally renowned publishers. His current research is focused on the area of conceptualization and development of biodegradable composites and cost-effective manufacturing methods. He was honored with the prestigious 'Excellent Research Contribution Award' of NIT Meghalaya in 2019 and 2023.

Fresh Frontier: Revolutionizing Perishable Supply Chain Through Digitization

Sneha Kushwaha and Saurabh Pratap

Abstract The following research implies the transformation of the perishable supply chain through digital processes, which use various digital technologies and are sustainable in handling the implementation of perishable products. A green food supply chain can be adapted by integrating environmentally responsible principles and conventions into supply chain management. Through research, various factors of barriers and drivers of the green food supply chain have been listed, which shows the impact on supply chain management. A poll-based survey was used to examine the factors; information was collected from various experts, and factor analysis was taken based on expert responses. The procedure of Fuzzy TOPSIS is used to identify the prioritized factors of barriers and drivers for the transformation of perishable products. By integrating digital technology into the supply chain, this research also aims to improve effective operations, reduce waste, and improve the performance for handling perishable goods. The study also provides an important perception for businesses seeking innovative solutions to improve their approach toward managing perishable goods effectively in today's dynamic industry.

Keywords Perishable · Green food supply chain · Fuzzy TOPSIS · Barriers · Drivers · Digitization

1 Introduction

Researchers have been interested in studying advanced ways to solve big problems. Nowadays, one main problem we encounter is the spread of fraudulent, misbranded, and unsafe food items, which can have harmful effects on health, and some products

S. Kushwaha
Department of Information Technology, Shivajirao S. Jondhale College of Engineering, Maharashtra, India

S. Pratap (✉)
School of Decision Science Engineering, Indian Institute of Technology BHU, Varanasi, India
e-mail: s.pratapiitkgp@gmail.com

M. Ram et al. (eds.), *Technological Advancements in Engineering and Management Sciences*, Management and Industrial Engineering,
https://doi.org/10.1007/978-3-032-04606-2_1

can also have destructive impacts on the environment (Kshetri 2023). As the Supply Chain plays an important role in the Business world which makes it easy to deal with consumer's needs. Due to the COVID-19 pandemic, many businesses had to adapt to the digital transformation, different changes in managing the Supply Chain networks and restrictions, and also remap the entire operational models, and find a way to sustain themselves in the unstable market (Vitsentzatou et al. 2022). The conversion of the common Supply Chain to a Digital Supply Chain had a great impact on Business management, as they have to adapt to the new technologies with different challenges. It involves different advanced technologies such as Internet of Things, data analytics, and blockchain technology for enhancing the effectiveness, transparency and sustainability of managing perishable goods from production to consumer. By digitizing the processes, businesses can optimize the processes related to storage, monitor the temperature control of the products, better transportation routes, ensure quality control measures, and make sure to meet the consumer demand for fresher products. Due to this advancement of digital technologies, Industry 4.0 also needs the Supply chain to embrace the advanced technologies and enhance their capabilities while transforming their business models and network setup to work together for vertical integration. This change is expected in common relationships between suppliers, dealers, retailers, manufacturers, and customers (Soosay and Kannusamy 2018). Customer behavior and expectations for the products are also changing due to digital transformation. To achieve these new expectations of customers, firms often need to enhance the traditional business models with online sales platforms such as directly delivering the products to buyers or select-and-collect options, which allow buyers to make purchases online and gather their products later from the shop (Subramanian et al. 2021).

2 Literature Review

The perishable goods industry faces different issues in managing the Supply Chains systematically and sustainably due to the nature of the products. Traditional supply chains often have some problems maintaining the quality of food, producing a lot of waste, and being unable to deliver the product on time, which is also a loss for a business. Nowadays many companies have started to adopt digital technologies to enhance efficiency, sustainability, transparency, and traceability, and due to this consumer demands are satisfied. The technologies that led the Supply Chain toward digitization are IoT, big data analytics, AI, and blockchain (Rejeb et al. 2022). There are many reasons that can be obstacles to the perishable supply chain. Through studying various research different drivers and barriers to the green food supply chain were studied. This includes barriers like cost considerations, lack of awareness which highlights the concern about the financial management in integrating digital tech in the supply chain, and lack of knowledge about the profitable gain in the businesses by integrating digitization can lead the business to face the loss (Kumar et al. 2020). Limitations of resources, technological limitations, and limited access to green technology also raise a concern about the lack of skilled workers, improper infrastructure, reducing the environmental impact and advanced eco-friendly technologies are the reasons for the disadvantages to sustainability (Guimarães 2022; Mehmood et al. 2021). The procedure of packaging is also one of the important activities in perishable food production. According to researchers, about 40% of food meant for people's use is discarded yearly. This waste is equal to nearly 10% of overall energy intake per person's weekly food intake (Verghese et al. 2015). The quality of perishable products and the quality of safely delivered products depend on packaging, hence it can also satisfy the consumers (Al-Ansari et al. 2020). Fulfilling the customer's demand, such as safe food products, transparency, and the delivery of the products, can attract more customers. Cost savings are an important factor for the growth of industries through the product supply chain and efficiency. There are some companies which have integrated digital technologies into their supply chain management for example Walmart has been using blockchain technology to improve the transparency and traceability of its food supply chain. Due to the use of blockchain technology, Walmart can monitor the movement of perishable goods from production to storage, improving food safety. However, adapting the new technologies with existing technologies needs careful planning and cooperation. Due to some challenges, the benefits of digitalization in the perishable supply chain are significant.

3 Methodology

The main objective of this study is to focus on the addition of a digital perishable supply chain into the business for profitable growth and customer satisfaction. Applying the Technique of Fuzzy TOPSIS to calculate the factors of barriers and drivers for the green food supply chain. Through literature review data is collected on the green food supply chain management, and surveys were conducted with the experts in the perishable goods, focusing on factors such as lack of awareness, technological limitations, employee engagement, environmental regulations and others. These factors were analyzed by applying the Fuzzy TOPSIS to identify the rank of barriers affecting the efforts in the supply chain and promoting the environmentally friendly operation in the perishable supply chains (Table 1).

The Fuzzy TOPSIS process is applied to find the ranks for prioritizing the factors for the perishable Supply Chain.

Table 1 Factors of Barriers and Drivers in green supply chain

Sr. no		Sub-factors	Description	References
1	Barriers	Cost considerations	It concerns about the financial impact of integrating the digital technologies into perishable Supply Chains, including both investment cost and ongoing operational expenses	Mehmood et al. (2021)
2		Lack of awareness	It refers to lack of understanding among the stakeholders about the benefits and impact of digitalization in the development of lost products	Govindan et al. (2014)
3		Limited of resources	This concerns about the limited financing, Shortage of skilled workers, and insufficient infrastructure which will affect the benefit of digital transformation projects	Sachin Kumar Mangla et al. (2018)
4		Technological limitations	It includes the issues like outdated infrastructure and unsuitable systems within the Supply Chain Network, which hinder the adoption of new technologies like IoT or AI	Kumar et al. (2021)
5		Complex supply chain	This highlights the problems caused by the complexity for handling the damaged products, with different stakeholders (such as farmers, processes) enabling the competition for effective coordination of activities using digital tools	Shisara et al. (2021)

(continued)

Table 1 (continued)

Sr. no		Sub-factors	Description	References
6		Short-term focus	Businesses often prioritize short-term profits over long-term safety and may not care about good environmental practices	Zkik et al. (2022)
7		Limited access to green technology	Limitation arises when companies lack access to advanced eco-friendly technologies, hindering optimized operations and reducing environmental impact in the food supply chain	Sachin Kumar Mangla et al. (2018)
8		Complex regulations	It highlights the problems, such as food safety, transportation, and data privacy, which may block the use of digital solutions	Mohseni et al. (2022)
9	Drivers	Consumer demand	It reflects growing customer demand for products that are safe, transparent, efficient, and deliverable	Kshetri (2023); Al-Ansari et al. (2020)
10		Cost saving	It reflects the potential for cost savings through process simplification through digital product management, and efficiency	Shisara et al. (2021)
11		Environmental concerns	It highlights the cost savings through digital product management and transportation optimization to increase efficiency	Mehmood et al. (2021)
12		Employee engagement	It focuses on using technology to increase employee engagement and productivity at various stages ensuring the smooth operations	Rejeb et al. (2022), Govindan et al. (2014)

(continued)

Table 1 (continued)

Sr. no		Sub-factors	Description	References
13		Supply chain resilience	It highlights the importance of making a strong network that can adapt quickly when disrupted by real-world data science	Vesela and Veleva (2019)
14		Supply chain transparency	It highlights the need to achieve consensus across the value chain using advanced traceability systems to ensure accountability at all levels	Subramanian et al. (2021), Kumar et al. (2021)
15		Environmental regulations	It recognizes the need to digitally implement sustainable practices to ensure compliance and promote green measures	Rejeb et al. (2022), Sachin Kumar Mangla et al. (2018)

3.1 Modified Fuzzy TOPSIS Procedure

Fuzzy TOPSIS or it is also called as Fuzzy Technique for Order Preference by Similarity to the ultimate Solution estimates the multiple alternatives against the selected factors. The method is built on the principle that the chosen alternatives must have the shortest distance to the Fuzzy Positive Ideal Solution (FPIS) and the farthest distance from the Fuzzy Negative Ideal Solution (FNIS) is chosen as optimal. A vertex method is used to calculate the distance between two triangular FNs. If $\tilde{o} = (x_1, y_1, z_1), \tilde{p} = (x_2, y_2, z_2)$ are two triangular FNs then:

$$d(\tilde{o}, \tilde{p}) = \sqrt{1/3[(x_1 - x_2)^2 + (y_1 - y_2)^2 + (z_1 - z_2)^2]} \tag{1}$$

Following steps involved in Fuzzy TOPSIS are as follows:

Step 1. Allot the ranking to the alternatives by Decision-makers.

Let's consider that there are two decision-maker groups with N members. The Fuzzy ranking of the nth decision-maker about the alternative A_i w.r.t. criterion C_j is indicated as $\tilde{a}_{ij}^n = (x_{ij}^n, y_{ij}^n, c_{ij}^n)$ and the weight criterion C_j is indicated as $\tilde{w}_j^n = (w_{j1}^n, w_{j2}^n, w_{j3}^n)$.

Step 2. Calculate the aggregated Fuzzy ranking of alternatives and the aggregated Fuzzy weights for criteria as shown in Table 2.

The aggregated fuzzy rating $\tilde{a}_{ij} = (x_{ij}, y_{ij}, z_{ij})$ of ith alternative w.r.t. jth criterion which is obtained as:

$$x_{ij} = \begin{matrix} \min \\ n \end{matrix} \left\{ x_{ij}^n \right\}, y_{ij} = \frac{1}{N} \sum_{N=1}^{N} y_{ij}^n,\ z_{ij} = \begin{matrix} \max \\ n \end{matrix} \left\{ c_{ij}^n \right\} \tag{2}$$

The total Fuzzy weight $\tilde{w}_j = (w_{j1}, w_{j2}, w_{j3})$ for the criterion C_j which are calculated by the formula:

$$w_{j1} = \begin{matrix} \min \\ n \end{matrix} \left\{ w_{j1}^n \right\},\ w_{j2} = \frac{1}{N} \sum_{N=1}^{N} w_{j2}^n,\ w_{j3} = \begin{matrix} \max \\ n \end{matrix} \left\{ w_{j3}^n \right\} \tag{3}$$

Table 2 Fuzzy linguistic scales

Linguistic terms	Triangular Fuzzy Number
Equally significant	1,1,3
Weakly significant	1,3,5
Moderately strong important	3,5,7
Very strongly important	5,7,9
Absolutely important	7,9,9

Step 3. Calculate the normalized fuzzy decision matrix.
The normalized Fuzzy decision matrix is $\overline{S} = [\overline{s}_{ij}]$, where

$$\overline{s}_{ij} = \left(\frac{x_{ij}}{z_j^*}, \frac{y_{ij}}{z_j^*}, \frac{z_{ij}}{z_j^*}\right) \textit{and } z_j^* = \underset{i}{\max}\left\{z_{ij}\right\}(\text{benefit criteria}) \tag{4}$$

Or

$$\overline{s}_{ij} = \left(\frac{x_j^-}{z_{ij}}, \frac{x_j^-}{y_{ij}}, \frac{x_j^-}{x_{ij}}\right) \textit{and } z_j^- = \underset{i}{\min}\left\{x_{ij}\right\}(\text{cost criteria}). \tag{5}$$

Step 4. Compute the weighted normalized Fuzzy decision matrix.
The weighted normalized Fuzzy decision matrix is $\tilde{V} = (\tilde{v}_{ij})$, where $\tilde{v}_{ij} = \tilde{s}_{ij} \times w_j$.

Step 5. Compute the Fuzzy Positive Ideal Solution (FPIS) and Fuzzy Negative Ideal Solution (FNIS).

The FPIS and FNIS are calculated as:

$$A^* = \left(\tilde{v}_1^*, \tilde{v}_2^*, \ldots, \tilde{v}_n^*\right), \text{ where } \tilde{v}_j^* = \underset{i}{\max}\left\{v_{ij3}\right\}; \tag{6}$$

$$A^- = \left(\tilde{v}_1^-, \tilde{v}_2^-, \ldots., \tilde{v}_n^-\right), \text{ where } \tilde{v}_j^- = \underset{i}{\min}\left\{v_{ij3}\right\} \tag{7}$$

Step 6. Compute the distance from each alternative to FPIS and FNIS.

$$\text{Let } d_i^* = \sum_{j=1}^{n} d(\tilde{v}_{ij}, \tilde{v}_j^*),\ d_i^- = \sum_{j=1}^{n} d(\tilde{v}_{ij}, \tilde{v}_j^*) \tag{8}$$

The distance from each alternative X_i to FPIS and FNIS, respectively.

Step 7. Calculate the closeness coefficient CC_i for each alternative.
Each alternative X_i we calculate the closeness coefficient CC_i as follows:

$$CC_i = \frac{d_i^-}{d_i^- + d_i^*} \tag{9}$$

Step 8. Rank the Category.

3.2 *Application of the Proposed Approach*

Step 1. Through literature review and the opinions of experts, 17 alternatives were selected.

Step 2. A questionnaire was distributed among the experts to get the interrelationships among the alternatives.

Step 3. Fuzzy TOPSIS process is used to compute the FPIS and FNIS and closeness coefficients.

Step 4. After calculating the comparison matrix, the Normalized Fuzzy decision matrix and weighted normalized Fuzzy decision matrix are computed and are shown in Tables 3 and 4.

Step 5. After calculating the normalized and weighted Fuzzy decision matrix FPIS and NPIS are computed and are shown in Tables 5 and 6.

Step 6. The further step is to compute the distance from each alternative to FPIS and NPIS, and is given in Table 7.

Steps 7 and 8. Compute closeness coefficients for each alternative and rank the alternatives as in Table 8.

4 Discussion

Through all the studies and research findings of the alternative data using different approaches, which involved the use of questionnaires, and also from existing literature review, and gathering the responses from various experts. Applying the Fuzzy TOPSIS process for alternatives for determining the priority from the given alternatives through finding the rank of each alternative, the rating of the various alternatives was determined by using this method. The results indicate the priority levels of different challenges and considerations that businesses must consider to achieve effective digitalization. Consumer demand is prioritized as one as its main focus is to deal with customers' needs, emphasizing the need for businesses to adapt quickly to consumer preferences and expectations. Cost savings and Supply Chain resilience are prioritized second and third, as cost saving is the important factor, which highlights the importance of financial management in the digitization process. Environmental concerns and Environmental regulations are prioritized as third and fourth, Navigating and complying with environmental regulations is a significant challenge that

Table 3 Weighted normalized decision matrix for the Barriers

	Barriers																							
	Cost consideration			Lack of awareness			Complex regulations			Limitation of resources			Technological supply chain			Complex supply chain			Short-term focus			Limited access to green tech		
A1	0.01	0.01	0.02	0.01	0.02	0.06	0.01	0.02	0.06	0.01	0.01	0.06	0.01	0.02	0.06	0.01	0.01	0.06	0.01	0.01	0.02	0.01	0.01	0.06
A2	0.01	0.02	0.06	0.01	0.01	0.06	0.01	0.02	0.06	0.01	0.01	0.01	0.01	0.01	0.02	0.01	0.01	0.01	0.01	0.02	0.06	0.01	0.01	0.02
A3	0.01	0.03	0.06	0.01	0.01	0.02	0.01	0.01	0.01	0.01	0.01	0.01	0.01	0.01	0.06	0.01	0.01	0.06	0.01	0.01	0.01	0.01	0.02	0.06
A4	0.01	0.02	0.06	0.01	0.01	0.06	0.01	0.01	0.02	0.01	0.02	0.06	0.01	0.01	0.06	0.01	0.01	0.01	0.01	0.02	0.06	0.01	0.01	0.06
A5	0.01	0.01	0.01	0.01	0.01	0.01	0.01	0.03	0.06	0.01	0.01	0.02	0.01	0.02	0.06	0.01	0.02	0.06	0.01	0.01	0.01	0.01	0.01	0.06
A6	0.01	0.01	0.06	0.01	0.01	0.06	0.01	0.02	0.06	0.01	0.03	0.06	0.01	0.02	0.06	0.01	0.02	0.06	0.01	0.01	0.02	0.01	0.01	0.06
A7	0.01	0.01	0.06	0.01	0.01	0.06	0.01	0.01	0.01	0.01	0.02	0.06	0.01	0.02	0.06	0.01	0.02	0.06	0.01	0.01	0.06	0.01	0.01	0.02
A8	0.01	0.03	0.06	0.01	0.02	0.06	0.01	0.01	0.06	0.01	0.01	0.02	0.01	0.02	0.06	0.02	0.06	0.06	0.01	0.01	0.06	0.01	0.02	0.06
A9	0.01	0.01	0.01	0.01	0.03	0.06	0.01	0.01	0.06	0.01	0.02	0.06	0.01	0.01	0.06	0.01	0.01	0.06	0.01	0.02	0.06	0.01	0.01	0.02
A10	0.01	0.01	0.02	0.01	0.02	0.06	0.01	0.02	0.06	0.01	0.02	0.06	0.01	0.01	0.06	0.01	0.01	0.06	0.01	0.01	0.02	0.01	0.02	0.06
A11	0.02	0.06	0.06	0.01	0.02	0.06	0.02	0.06	0.06	0.01	0.01	0.01	0.01	0.01	0.01	0.01	0.03	0.06	0.01	0.01	0.02	0.01	0.01	0.01
A12	0.01	0.03	0.06	0.01	0.02	0.06	0.01	0.02	0.06	0.01	0.02	0.06	0.01	0.01	0.06	0.01	0.02	0.06	0.01	0.02	0.06	0.01	0.01	0.06
A13	0.01	0.01	0.02	0.01	0.02	0.06	0.01	0.01	0.02	0.01	0.02	0.06	0.01	0.01	0.01	0.01	0.01	0.06	0.02	0.06	0.06	0.01	0.01	0.02
A14	0.01	0.03	0.06	0.01	0.02	0.06	0.01	0.02	0.06	0.01	0.01	0.02	0.01	0.01	0.06	0.01	0.01	0.02	0.01	0.02	0.06	0.01	0.02	0.06
A15	0.01	0.02	0.06	0.02	0.03	0.06	0.01	0.02	0.06	0.01	0.02	0.06	0.02	0.03	0.06	0.01	0.01	0.06	0.03	0.04	0.06	0.01	0.01	0.06

Table 4 Weighted normalized decision matrix for the Drivers

Drivers																				
Cost saving			Environmental concerns			Employee engagement			Supply chain resilience			Supply chain transparency			Environmental regulations			Consumer demand		
0.01	0.03	0.06	0.01	0.03	0.06	0.01	0.03	0.06	0.01	0.01	0.02	0.01	0.03	0.05	0.01	0.03	0.05	0.03	0.04	0.05
0.01	0.04	0.06	0.02	0.04	0.06	0.01	0.04	0.06	0.01	0.03	0.05	0.03	0.05	0.06	0.03	0.05	0.06	0.02	0.04	0.05
0.01	0.04	0.06	0.01	0.03	0.05	0.01	0.03	0.05	0.01	0.03	0.06	0.01	0.04	0.06	0.03	0.05	0.06	0.01	0.02	0.03
0.03	0.05	0.06	0.01	0.01	0.03	0.03	0.05	0.06	0.01	0.03	0.06	0.01	0.03	0.06	0.01	0.02	0.03	0.01	0.01	0.02
0.01	0.03	0.05	0.02	0.04	0.06	0.01	0.03	0.06	0.02	0.03	0.05	0.03	0.05	0.06	0.03	0.05	0.06	0.03	0.04	0.05
0.03	0.05	0.06	0.02	0.04	0.06	0.01	0.01	0.03	0.01	0.03	0.06	0.03	0.05	0.06	0.01	0.03	0.06	0.01	0.02	0.04
0.03	0.05	0.06	0.01	0.03	0.06	0.03	0.05	0.06	0.01	0.04	0.06	0.01	0.03	0.05	0.01	0.03	0.06	0.01	0.02	0.05
0.01	0.04	0.06	0.01	0.02	0.05	0.02	0.04	0.06	0.01	0.01	0.03	0.01	0.02	0.05	0.01	0.01	0.03	0.01	0.02	0.03
0.01	0.03	0.06	0.01	0.03	0.05	0.01	0.03	0.06	0.02	0.03	0.05	0.01	0.02	0.03	0.01	0.03	0.06	0.04	0.05	0.05
0.03	0.05	0.06	0.03	0.05	0.06	0.03	0.05	0.06	0.01	0.01	0.02	0.01	0.03	0.06	0.01	0.03	0.05	0.01	0.02	0.03
0.02	0.04	0.06	0.02	0.05	0.06	0.01	0.03	0.05	0.01	0.02	0.03	0.01	0.01	0.03	0.02	0.04	0.06	0.01	0.01	0.02
0.01	0.04	0.06	0.01	0.03	0.06	0.01	0.03	0.06	0.02	0.03	0.05	0.02	0.04	0.06	0.01	0.03	0.05	0.03	0.04	0.05
0.01	0.01	0.03	0.02	0.03	0.05	0.01	0.03	0.06	0.02	0.04	0.06	0.03	0.05	0.06	0.01	0.01	0.02	0.02	0.03	0.04
0.02	0.04	0.06	0.01	0.03	0.06	0.03	0.05	0.06	0.01	0.03	0.06	0.01	0.04	0.06	0.01	0.02	0.05	0.01	0.01	0.03
0.01	0.03	0.06	0.01	0.03	0.06	0.02	0.05	0.06	0.01	0.04	0.06	0.01	0.04	0.06	0.01	0.03	0.06	0.01	0.03	0.05

Table 5 Distance of FPIS

Distance from FPIS															
A1	0.05	0.04	0.04	0.04	0.04	0.04	0.05	0.04	0.04	0.03	0.04	0.05	0.04	0.04	0.01
A2	0.04	0.04	0.04	0.01	0.01	0.05	0.04	0.04	0.04	0.04	0.03	0.04	0.04	0.04	0.02
A3	0.03	0.01	0.01	0.01	0.04	0.04	0.05	0.04	0.03	0.03	0.04	0.04	0.04	0.05	0.02
A4	0.04	0.04	0.01	0.04	0.04	0.05	0.04	0.05	0.02	0.05	0.04	0.04	0.04	0.02	0.01
A5	0.05	0.01	0.04	0.01	0.04	0.04	0.05	0.03	0.04	0.04	0.04	0.05	0.05	0.05	0.04
A6	0.04	0.04	0.04	0.04	0.04	0.04	0.05	0.05	0.04	0.02	0.04	0.05	0.05	0.04	0.03
A7	0.04	0.04	0.01	0.04	0.04	0.04	0.04	0.05	0.04	0.05	0.04	0.03	0.03	0.04	0.03
A8	0.03	0.04	0.04	0.05	0.04	0.02	0.04	0.04	0.03	0.04	0.02	0.03	0.03	0.02	0.02
A9	0.05	0.04	0.04	0.04	0.04	0.04	0.04	0.04	0.03	0.04	0.04	0.02	0.02	0.04	0.05
A10	0.05	0.04	0.04	0.04	0.04	0.04	0.05	0.05	0.05	0.05	0.01	0.04	0.04	0.03	0.02
A11	0.02	0.04	0.05	0.05	0.05	0.03	0.05	0.04	0.05	0.03	0.02	0.02	0.02	0.04	0.01
A12	0.03	0.04	0.04	0.04	0.04	0.04	0.04	0.04	0.04	0.04	0.04	0.04	0.04	0.03	0.04
A13	0.05	0.04	0.01	0.04	0.05	0.04	0.02	0.02	0.04	0.04	0.04	0.05	0.05	0.01	0.03
A14	0.03	0.04	0.04	0.05	0.04	0.05	0.04	0.04	0.04	0.05	0.04	0.04	0.04	0.03	0.02
A15	0.04	0.04	0.04	0.04	0.03	0.04	0.04	0.04	0.04	0.05	0.04	0.04	0.04	0.04	0.03

demands to be both compliant and environmentally responsible. Addressing environmental concerns has become more important, which is impacting the reputation and efforts of businesses. Supply Chain Transparency and Employee engagement are ranked sixth and seventh. Receiving Supply Chain transparency ensures a piece clear information about sourcing and production, building consumers trust. Engaging the employees in the digitization process is important, making sure that they are well-informed and invested in the new technologies and processes, these are the Drivers which were prioritized at the top compared to the barriers of the food supply chain which would be helpful for the transformation of the digitization in Supply chain management. While the barriers which can affect digitization are prioritized at the lowest, the lack of awareness and understanding the digital technologies is important for the adoption and integration of digital technologies into Supply Chain operations. Limited access to green technology, and technology limitations are the factors which can impact business growth, as well as impact on the environment. Complex regulations further highlight the regulatory complexities affecting operational efficiency. The result provides valuable insights for prioritizing the actions that aim to enhance the perishable Supply Chain through digital transformation while emphasizing key challenges across multiple dimensions.

Table 6 Distance of NPIS

Distance from FNIS																
A1	0.01	0.03	0.03	0.03	0.03	0.03	0.01	0.01	0.03	0.03	0.03	0.03	0.01	0.03	0.03	0.03
A2	0.03	0.03	0.03	0.00	0.01	0.00	0.04	0.01	0.04	0.04	0.04	0.04	0.03	0.05	0.05	0.04
A3	0.03	0.01	0.00	0.00	0.03	0.03	0.00	0.04	0.04	0.03	0.03	0.03	0.04	0.04	0.05	0.02
A4	0.03	0.03	0.01	0.03	0.03	0.00	0.03	0.04	0.05	0.02	0.05	0.02	0.04	0.04	0.02	0.01
A5	0.05	0.00	0.03	0.01	0.03	0.03	0.00	0.04	0.03	0.04	0.04	0.04	0.04	0.05	0.05	0.04
A6	0.03	0.03	0.03	0.03	0.03	0.03	0.01	0.04	0.05	0.04	0.02	0.04	0.04	0.05	0.04	0.03
A7	0.03	0.03	0.00	0.03	0.03	0.03	0.03	0.01	0.05	0.04	0.05	0.04	0.04	0.03	0.04	0.03
A8	0.03	0.03	0.03	0.05	0.03	0.04	0.03	0.04	0.04	0.03	0.04	0.03	0.02	0.03	0.02	0.02
A9	0.00	0.03	0.03	0.04	0.03	0.03	0.03	0.01	0.04	0.03	0.04	0.03	0.04	0.02	0.04	0.05
A10	0.01	0.03	0.03	0.03	0.03	0.03	0.01	0.04	0.05	0.05	0.05	0.05	0.01	0.04	0.03	0.02
A11	0.04	0.03	0.04	0.00	0.00	0.03	0.01	0.01	0.04	0.05	0.03	0.05	0.02	0.02	0.04	0.01
A12	0.03	0.03	0.03	0.03	0.03	0.03	0.03	0.04	0.04	0.04	0.04	0.04	0.04	0.04	0.03	0.04
A13	0.01	0.03	0.01	0.03	0.00	0.03	0.04	0.01	0.02	0.04	0.04	0.04	0.04	0.05	0.01	0.03
A14	0.03	0.03	0.03	0.01	0.03	0.01	0.03	0.04	0.04	0.04	0.05	0.04	0.04	0.04	0.03	0.02
A15	0.03	0.03	0.03	0.03	0.03	0.03	0.04	0.04	0.04	0.04	0.05	0.04	0.04	0.04	0.04	0.03

Table 7 Distance of FPIS and NPIS (d*and d–)

	A1	A2	A3	A4	A5	A6	A7	A8	A9	A10	A11	A12	A13	A14	A15
d*	0.58	0.49	0.49	0.53	0.57	0.59	0.55	0.51	0.56	0.59	0.53	0.57	0.51	0.59	0.57
d–	0.40	0.46	0.41	0.44	0.49	0.51	0.49	0.50	0.47	0.46	0.40	0.53	0.41	0.48	0.55

Table 8 Closeness Coefficients and Ranking of alternatives

CCi	0.50	0.49	0.48	0.48	0.47	0.46	0.46
Factors	Environmental concerns	Consumer demand	Supply chain resilience	Lack of awareness	Supply chain transparency	Environmental regulations	cost saving
	A8	A15	A12	A2	A7	A6	A5
Rank	1	2	3	4	5	6	7

0.46	0.46	0.46	0.45	0.45	0.44	0.43	0.41
Cost considerations	Technological limitations	Employee engagement	Complex supply chain	Limited access to green tech	Short-term focus	Complex regulations	Limitation of resources
A4	A9	A3	A14	A13	A10	A11	A1
8	9	10	11	12	13	14	15

5 Conclusion

This study highlights the important factors that lead the digital transformation of perishable Supply Chains through Fuzzy TOPSIS analysis. Consumer demand and cost savings are prioritized at the top as they highlight the importance of market operations and financial efficiency as an important driver for growth. Adopting the supply chain resilience in business transformation is also important, and following the government regulations can indicate the development of business management. Overcoming challenges like limited access to green technology and technological limitations is crucial for effectively using digital technology. Short-term focus and limitation of resources are the problems that highlight the need for collaborative plans to improve operational efficiency. This research highlights the value of adopting strategies that include cost control, regulatory compliance, and advanced technology. With these strategies, businesses can transform the perishable supply chain through digitalization and guarantee their long-term sustainability and success.

References

A. C. F. C. Yana Medeiros Guimarães, João Henrique Paulino Pires Eustachio, Walter Leal Filho, Letícia Franco Martinez, Maurício Ribeiro do Valle, "Drivers and barriers in sustainable supply chains: The case of the Brazilian coffee industry, Sustainable Production and Consumption.," vol. 34, pp. 42–54, 2022, [Online]. Available: https://doi.org/10.1016/j.spc.2022.08.031.%0A

T. Al-Ansari, M. Haji, L. Kerbache, and M. Muhammad, "Roles of Technology in Improving Perishable Food Supply Chains," *Logistics*, vol. 4, no. 4, p. 33, 2020.

B. W. C. Vesela R. Veleva, "The role of drivers, barriers, and opportunities of green chemistry adoption in the major world markets, Current Opinion in Green and Sustainable Chemistry," vol. 19, pp. 30–36, 2019, [Online]. Available: https://doi.org/10.1016/j.cogsc.2019.05.001.%0A

R. P. C. Calafat-Marzal, M. Sánchez-García, L. Marti, "Agri-food 4.0: Drivers and links to innovation and eco-innovation, Computers and Electronics in Agriculture.," vol. 207, 2023.

K. Govindan, M. Kaliyan, D. Kannan, and A. N. Haq, "Barriers analysis for green supply chain management implementation in Indian industries using analytic hierarchy process," *Int. J. Prod. Econ.*, vol. 147, no. PART B, pp. 555–568, 2014, https://doi.org/10.1016/j.ijpe.2013.08.018.

M. Javaid, A. Haleem, R. P. Singh, S. Rab, R. Suman, and I. H. Khan, "Evolutionary trends in progressive cloud computing based healthcare: Ideas, enablers, and barriers," *Int. J. Cogn. Comput. Eng.*, vol. 3, no. February, pp. 124–135, 2022, https://doi.org/10.1016/j.ijcce.2022.06.001.

N. Kshetri, "Blockchain's Role in Enhancing Quality and Safety and Promoting Sustainability in the Food and Beverage Industry," *Sustainability*, vol. 15, no. 23, p. 16223, 2023, https://doi.org/10.3390/su152316223.

S. Kumar, A., Mangla, S. K., Kumar, P., & Karamperidis, "Challenges in perishable food supply chains for sustainability management: A developing economy perspective. Business Strategy and the Environment.," 2020, [Online]. Available: https://doi.org/10.1002/bse.2470

S. Kumar, R. D. Raut, K. Nayal, S. Kraus, V. S. Yadav, and B. E. Narkhede, "To identify industry 4.0 and circular economy adoption barriers in the agriculture supply chain by using ISM-ANP," *J. Clean. Prod.*, vol. 293, p. 126023, 2021, https://doi.org/10.1016/j.jclepro.2021.126023.

A. Mehmood, S. Ahmed, E. Viza, A. Bogush, and R. M. Ayyub, "Drivers and barriers towards circular economy in agri-food supply chain: A review," *Bus. Strateg. Dev.*, vol. 4, no. 4, pp. 465–481, 2021, https://doi.org/10.1002/bsd2.171.

S. Mohseni, K. Baghizadeh, and J. Pahl, "Evaluating Barriers and Drivers to Sustainable Food Supply Chains," *Math. Probl. Eng.*, vol. 2022, 2022, https://doi.org/10.1155/2022/4486132.

A. Rejeb, K. Rejeb, A. Abdollahi, S. Zailani, M. Iranmanesh, and M. Ghobakhloo, "Digitalization in food supply chains: A bibliometric review and key-route main path analysis," *Sustain.*, vol. 14, no. 1, pp. 1–29, 2022, https://doi.org/10.3390/su14010083.

M. A. K. K. Shisara, Y. N. S. Wijewardana, and R. A. P. I. S. Dharmadasa, "Barriers and Drivers to Green Supply Chain Management Practices: Evidence from Manufacturing and Agricultural Sectors of Sri Lanka," *Asian J. Sci. Appl. Technol.*, vol. 10, no. 2, pp. 5–14, 2021, https://doi.org/10.51983/ajsat-2021.10.2.3038.

P. J. Silva, V., Sereno, A.M. & do Amaral Sobral, "Food Industry and Processing Technology: On Time to Harmonize Technology and Social Drivers.," pp. 1–13, 2018, [Online]. Available: https://doi.org/10.1007/s12393-017-9164-8

C. Soosay and R. Kannusamy, "Scope for Industry 4.0 in Agri-food Supply Chain," *Hambg. Int. Conf. Logist. 2018*, pp. 37–56, 2018, [Online]. Available: https://tore.tuhh.de/handle/11420/1787

A. Subramanian, K. J. Yermal, S. Hemaraju, and S. Ananth, "Effectiveness of Digitalized Supply Chain During 2020 COVID-19 Pandemic: Case Studies," *Int. J. Res. Eng. Sci. Manag.*, vol. 4, no. 6, pp. 39–43, 2021.

R. et al. Testa, F., Pretner, G., Iovino, "Drivers to green consumption: a systematic review.," 2021, [Online]. Available: https://doi.org/10.1007/s10668-020-00844-5

H. Verghese, K., Lewis, H., Lockrey, S., & Williams, "Packaging's Role in Minimizing Food Loss and Waste Across the Supply Chain.," pp. 603–620, 2015, [Online]. Available: https://doi.org/10.1002/pts.2127

E. Vitsentzatou, G. T. Tsoulfas, and A. N. Mihiotis, "The Digital Transformation of the Marketing Mix in the Food and Beverage Service Supply Chain: A Grey DEMATEL Approach," *Sustain.*, vol. 14, no. 22, 2022, https://doi.org/10.3390/su142215228.

Y. K. D. Sachin Kumar Mangla, Sunil Luthra, Nick Rich, Divesh Kumar, Nripendra P. Rana, "Enablers to implement sustainable initiatives in agri-food supply chains.," *Int. J. Prod. Econ.*, vol. 203, pp. 379–393, 2018.

F. E. Zkik, K., Belhadi, A., Rehman Khan, S. A., Kamble, S. S., Oudani, M., & Touriki, "Exploration of barriers and enablers of blockchain adoption for sustainable performance: implications for e-enabled agriculture supply chains. International Journal of Logistics Research and Applications," *Int. J. Logist. Res. Appl.*, 2022.

Production Planning and Resource Optimization by Using MRP in MSMEs: A Case Study of an Indian Gear Manufacturer

M. Madhu Chauhan, Sonu Rajak, Prakash Kumar, and Chandan Pandey

Abstract Production planning is very important for the success of any manufacturing organization. To gain a competitive advantage in micro and small enterprises (MSMEs), it is important to utilize a robust production planning and resource optimization tool. This chapter studies production planning, such as scheduling, inventory control, and optimization, through adopting material requirements planning (MRP). The aim of the study is to study the efficiency and effectiveness of MRP. The study has been conducted in an Indian gear manufacturing organization. The result shows that the MPR enhances tracking and product visibility and generates scheduling to meet the daily customer demand. The MRP effectively generates the planned order release. The chapter also suggests to the organization how to implement MRP and use it to its advantage in its production planning process.

Keywords Production planning · MRP · Gear manufacturer · Competitive advantage

M. Madhu Chauhan · P. Kumar (✉)
Department of Production Engineering, National Institute of Technology Tiruchirappalli, Tiruchirappalli, India
e-mail: prakashkumar@nitt.edu

S. Rajak
Department of Mechanical Engineering, National Institute of Technology Patna, Patna, India

C. Pandey
Department of Mechanical Engineering, Indian Institute of Technology Jodhpur, Jodhpur, India

M. Ram et al. (eds.), *Technological Advancements in Engineering and Management Sciences*, Management and Industrial Engineering,
https://doi.org/10.1007/978-3-032-04606-2_2

1 Introduction

1.1 Industry Context

The gear manufacturing industry plays a crucial role in various sectors, including automotive, aerospace, and machinery. This chapter explores the implementation of MRP in a gear plant to address challenges in inventory management, production scheduling, and resource allocation. This chapter attempts to suggest practical methods for successfully deploying MRP systems in a gear plant context by analyzing the difficulties faced by gear plants and analyzing current MRP practices. The study aimed to address the need for implementing MRP in gear plants, offering insights and recommendations for improving operational efficiency and minimizing production costs. Currently, the gear plant generates production orders manually. To solve this, we tried to implement MRP, and in the upcoming part of that gear plant produced, Product specifications, Plant's layout, an introduction about MRP, SAP implementation, and its integration in the upcoming lines. Gear plant produces hypoid gear set of different models. The description of the hypoid gear set is as follows.

Hypoid gear: A hypoid gear set is a type of gear system commonly used in mechanical power transmission applications. It consists of two gears: a pinion gear and a ring gear. Unlike a traditional spiral bevel gear set, which has intersecting axes of rotation, a hypoid gear set has non-intersecting and offset axes of rotation.

Applications: Hypoid gear sets find application in various industries, including automotive, industrial machinery, and household appliances. They are commonly used in rear-wheel-drive vehicles, where they enable the transmission of power from the engine to the rear axle.

There are various hypoid gear variants available, and each has a unique design and use. Several popular hypoid gear models are, such as M160 3.73, M160 2.53, M180 4.1, M180 4.3, M186 4.1, M205 4.56, M216, and 3.73. The Gear—hypoid gear set M184 4.1 is shown in Fig. 1.

Where M—indicates in mm, 3.73 in (M160 3.73) indicates the gear ratio i.e., ratio of the number of teeth in the gear to the number of teeth in the pinion. In Fig. 1,

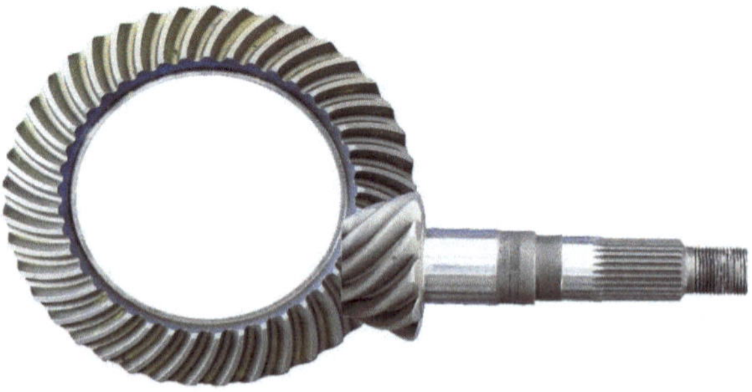

Fig. 1 Gear-hypoid gear set M184 4.1

the Number of teeth on the ring gear is 41; the number of teeth on the pinion is 10. Therefore, the gear ratio is 4.1.

1.2 Material Requirements Planning

MRP, which stands for material requirements planning, is a methodology and system used by organizations to effectively plan and manage their material resources for production and inventory control. MRP helps ensure that the right materials are available at the right time and in the right quantities to meet production needs and customer demands.

SAP MRP: SAP MRP is a module within the SAP ERP (Enterprise resource planning) system that helps organizations manage their materials and inventory planning processes.

a. Demand management
b. Bill of materials
c. Master data
d. Planning parameters
e. MRP Run
f. Procurement and Production Integration
g. Exception Monitoring
h. Reporting and Analytics.

The SAP production planning and control process relies heavily on planned orders. To meet consumer demand and maximize their manufacturing processes, they help organizations in effectively managing their material requirements and production schedules. The gear plant's planned MRP implementation methodology takes a methodical approach to guarantee accurate and efficient planning. The MRP process's foundation is set by updating the master data, allowing the system to operate with up-to-date, reliable information. Existing planned orders, inbound deliveries, schedule lines, purchase orders, and purchase requisitions can all be deleted to get rid of any outdated or unnecessary information that might affect the MRP calculations. The MRP process is started by running the T code MD02, which results in the creation of new planned orders that are in line with the most recent requirements and data. With this strategy, the gear plant hopes to increase the effectiveness of planning and scheduling overall and streamline the MRP implementation.

Further in Sect. 2, a brief explanation of the literature review, gaps have been identified. Later, in Sect. 3, how the gear plant works, its processes, and the approach of MRP through SAP are explained. In Sect. 4, planned order results have been discussed and explained that MRP will be a continuous improvement project. In Sect. 5, it is concluded that through SAP MRP, inventory management will be done, and planned orders will be generated automatically. Later, we discussed how and where SAP integration helps the manufacturing industries to survive in the market, and how MRP will be integrated with the machine learning and artificial intelligence.

2 Literature Review

The following section presents a literature review focused on the implementation of SAP MRP in gear plants. This review aims to provide a comprehensive understanding of the current state of research and practice in gear plants, highlighting key findings, methodologies, and trends. This study looks at the available literature to find gaps, problems, and opportunities in the deployment of SAP MRP systems designed specifically for gear manufacturers.

Malindzakova et al. (2022) proposed that the ABC inventory analysis method from lean management and the MRP approach are combined into a logical framework. A case study firm in engineering production significantly decreased changeover downtimes by using this combination method. In order to improve the fundamental supply chain management (SCM) ideas of inventory control and forecasting in a distribution network, Angolia and Pagliari (2018) used SAP-ERP-based procurement strategies and logistics plans. MRP was employed by Ardiyanto et al. (2022) to increase production cost efficiency in PT Petrokimia Gresik, a fertilizer production company. MRP was used by Atmika et al. (2022) to maximize inventory for MSMEs. It enables the company to effectively acquire materials in order to satisfy customer demand. In order to create raw material ordering planning systems for the apparel industry, Hasanati et al. (2018) use PHP programming and FIREBIRD database-based MRP. The production schedule generated by the production scheduling information system is fed into this system, which then outputs suggestions for scheduling the orders of raw materials. In order to forecast the inventory's backorder, Kumar et al. (2025) used a robust machine learning algorithm, such as XGBoost and random forest. The level of stockout, overstock, and non-moving stock is used to gauge the performance of the inventory. In order to manage inventory in a plastic injection manufacturing company, Hartoyo and Arvitrida (2023) implemented a hybrid MRP-Just-In-Time (JIT) system. In terms of planning, MRP is used to enhance the inventory process, whereas JIT is used to expedite the inventory process in terms of direct field operational implementation. Rajak et al. (2025) examine the obstacles that manufacturing organizations face when attempting to integrate Lean Six Sigma with Industry 4.0. Kasat and Deshmukh (2020) employ MRP, Equator, and S&OP to rectify the production planning problems encountered and enhance productivity. Its goal is to complete the gap between the demand side (customers) and the supply side (Company A). In order to support closed-loop planning and address complaints about rigid, arbitrary, and erratic vendors, it is necessary to design and implement business processes. This fosters competition to become more affordable, quicker, dependable, and adaptable. The company can gain a competitive edge on the job market and is exposed to industry-standard tools and practices that inclusion of SAP ERP as a commercial software application in the simulation (Jessica et al. 2022). Sato et al. (2016)

explored the implementation of distribution BoM (D-BoM) in SAP R/3 for whole supply chain planning with MRP. By adapting D-BOM and utilizing related routings and work centers, the study demonstrates the suitability of MRP for integrated supply chain planning. The integration of D-BOM and manufacturing BOM in SAP ERP proves effective, leveraging the MRP function to provide planning information for distribution and production processes. Future research directions include improving planning operations in ERP and applying throughput control to dynamic supply chain management.

3 Methodology

To ensure the successful and efficient production of gears, a gear plant must be designed with several factors and stages in mind. The main elements of gear plant design are summarized in Fig. 2.

Soft cell: Pinion outer diameter turning, spline rolling, hobbing, chamfering will be done here. Ring gear outer diameter turning, hobbing will be done here.

Heat Treatment: In the crucial process of heat treatment gears, which involves heating them to high temperatures and then quickly cooling them, the molecular structure of the material is changed, increasing the hardness of the gears. Gears with heat treatment provide stronger mechanical characteristics.

Hard cell: Turning of outer diameter for both pinion and ring gear will be done, coating of lapping paste, and chamfering will be done.

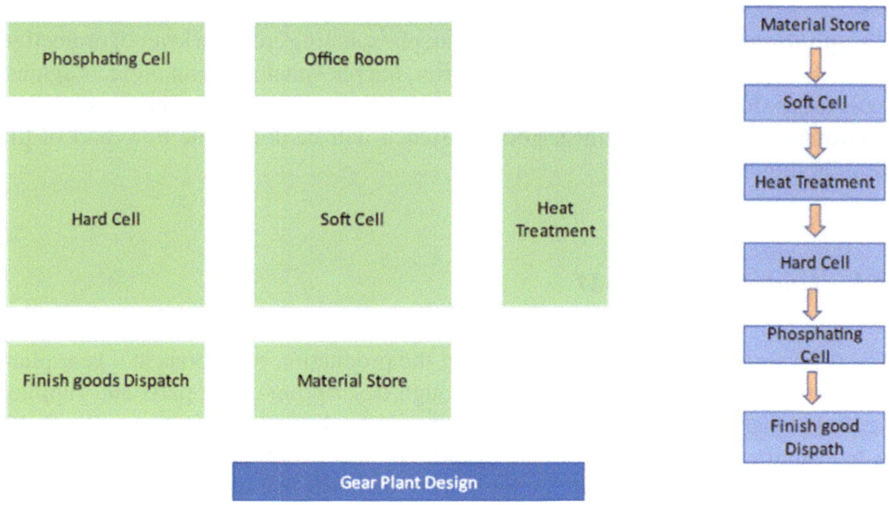

Fig. 2 Layout of gear plant

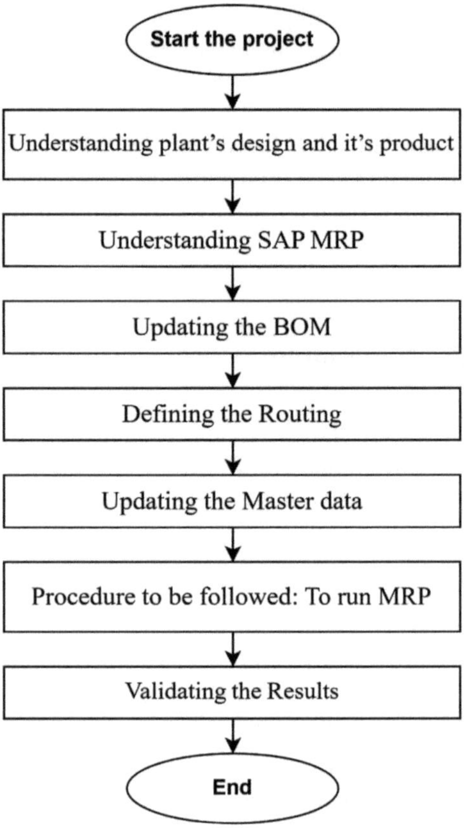

Fig. 3 Flowchart of the work

Phosphating cell: On the surface of the gear, phosphating creates a layer of protection that aids in preventing corrosion. As a barrier, the phosphate coating protects against moisture, oxidation, and other corrosive substances that are present in the operational environment. Finally, the finish goods dispatch will be done. The flowchart of the work is shown in Fig. 3.

3.1 Updating the BoM

The most important aspect for monitoring the production operations at a gear plant is updating the BoM. The BoM is a thorough description of the parts and supplies needed to make a gear or assembly. Those parts have been added to the regular BoM. A few hundred gear sets will be sent outside for heat treatment due to plant capacity restrictions. The normal BoM did not include the parts that will be sent out.

3.2 Defining the Routing

In manufacturing and production management, defining the routing also referred to as the routing sheet or process plan, is an essential stage. It describes the series of procedures and workstations necessary to convert raw materials or components into final goods. Identify the characteristics, requirements, and manufacturing specifications of the product. Determine the essential parts, subassemblies, and procedures entailed in the manufacturing process. Split the manufacturing process into several operations or steps needed to convert the raw materials or component parts into the finished product. Determine the proper order of processes by considering elements including technological needs, efficiency, and logical flow. Identify the production or workstation locations where each procedure will be carried out. Consider elements like the equipment's capacity, capabilities, and any specialized equipment or tools needed for jobs. Determine the resources needed for each operation, including the people, tools, machinery, fixtures, and any other supplies or equipment needed to finish the job. Give detailed directions, measurements, tasks, quality standards, and any additional considerations or safety measures for each operation. The routing's key quality control points should be identified. To guarantee that the product satisfies the necessary quality standards, decide where inspections, testing, or quality checks should be conducted. Make a thorough routing document or sheet that includes the data defined in the previous phases. The operation sequence, workstations, resources, operation specifics, time estimates, and quality control points should all be included in this document. Make sure the production team can readily understand the routing and that it is clear and organized. Distribute the routing document to all pertinent parties, including the operators, managers, and other production-related employees. Update the routing in SAP MRP.

3.3 Updating the Master Data

To ensure accurate and efficient planning and execution of manufacturing and procurement operations, SAP MRP master data must be updated. Data on the materials used in production is contained in the material master record. Access the material master module in SAP and modify fields like the material description, unit of measurement, procurement type, planning parameters (safety stock, lot size, reorder point etc.,), and the BoM, if necessary, to update the material master data. Make sure the data appropriately reflects the qualities and demands of the subject. If there are any changes to the work center data, go to the SAP work center module and make the necessary updates to the work center records. Change any descriptions, capacities, production schedules, or other factors that have an impact on capacity planning and scheduling for work centers. The parameters for planning and scheduling material requirements include lead time, safety stock, lot size, and planning horizon. Based on the needs and properties of the material, access SAP's MRP module and update the material-specific MRP parameters. To make sure that the revised data works as intended and has no unforeseen effects, thoroughly test and validate the changes in a test environment before implementing them. Table 1 shows the T codes and their description.

The T codes (Transaction code), SAP works based on T codes. These are the T codes we have used for this study. Internally in SAP, T codes are present; we need to integrate them and use them for our study purposes.

T Code: MM02 [To change Master data].

T Code: MM03 [Display Master Data/MRP1, MRP2, MRP3, MRP4].

Update Master data

MRP 1: Changes to be made in MRP1 for finished goods are shown in Table 2.

Table 2 consists of variables which we need to update in SAP, based on plant capacity and product type. And, Changes to be made in MRP1 for child parts are shown in Table 3.

Table 3 describes that for child parts, only Lot size and MRP type should be changed here.

MRP 2: Changes to be made in MRP 2 for finished goods are shown in Table 4.

Table 1 T codes and their description

T Code	Description
MM02	Change master data
MM03	Display MRP
ZPPR15	BOM display
MD16	Delete planned orders
ZINBOUND_DEL	Deleting inbound deliveries
MD04	Delete schedule lines, Purchase Requisition, Purchase orders (po item)
COOIS AND COHV	Production order deletion
MD02	Rum MRP

Table 2 Changes to be made in MRP1 for finished goods

Finished good	
MRP type	VB (Manual reorder planning)
Reorder point	#value#
MRP controller	Z04
Lot size	ZC
Min. lot size	#value#
Max. lot size	#value#
Max. stock level	#value#

Table 3 Changes to be made in MRP1 for child parts

Child parts	
MRP type	PD
Lot size	ZX (SIL: lot for lot quantity)

Table 4 Changes to be made in MRP 2 for finished goods

Finished good	
Procurement type	E
Storage location	#value#
Safety stock	#value#

Table 5 Changes to be made in MRP 2 for child parts

Child parts	
Procurement type	E/X/F
Storage location	#value#
Safety stock	#value#

Table 4 describes the data we need to update in MRP 2, where the product should go after finishing a certain process that needs to be defined here. And, Changes to be made in MRP2 for child parts are shown in Table 5.

Table 5 describes that for child parts, which have procurement type, need to be updated based on available data, and also storage location. We do not disturb much of MRP 1, 2 for child parts as it should be kept with zero inventory. Transfer the updated master data to the production environment when all the changes have been completely tested. To use the revised master data for MRP planning and execution, activate it. It is essential that you stick to appropriate change management processes and record the modifications made.

3.4 BoM Display

T code: ZPPR15

IN SAP, Go to ZPPR15 to display the Bill of Materials. Take the finished part number, and then Go to T code ZPPR15, then enter, plant number, the finished part number, multi-level, and after that, execute. So, after entering T code ZPPR15, the plant number needs to be entered along with the finish part number by keeping multi-level BoM. So, after entering into ZPPR15, BOM will be displayed and kept in a running tab that should be used in the upcoming steps to understand the finished part and its child parts based on BOM.

3.4.1 Deleting Planned Orders

T code: MD16

Take the finished part/Child part number from BoM, then Go to MD16, after that Select material, then enter. By entering the T code MD16, note down the finished part number and select the material option form here, and then execute. Then, planned orders will be displayed in the next step. Enter the finished part number, then execute. Here, enter the material number and plant, then execute. Planned orders will be displayed in the diagonal box, then delete all those planned orders with the help of the delete option. After entering into planned orders T code, now select the planned orders if present, then select the delete option such that non-planned orders should be deleted for both finished part and child parts by following the same procedure.

3.4.2 Deleting Inbound Deliveries

T code: zinbound_del

Enter child parts of having procurement type x and f, copy here in material number, then enter plant number, then keep goods movement type a to b, then after executing, delete those inbound deliveries. T code Zinbound_del, here enter the material number, goods movement type, and plant number, then execute. Here, we need to make all inbound deliveries and delete all the inbound deliveries you see here, then save and exit.

3.4.3 Deleting Scheduled Lines

T Code: MD04

After entering T code: MD04, enter finished part number, child part's number, plant number then delete all the scheduled lines, purchase requisition, purchase orders then here whatever the scheduled lines you see here delete then, also delete purchase orders, purchase requisition etc., keep it clean for the MRP. Now, here just make sure that all the PR's, PO's, and scheduled lines have been deleted.

3.4.4 Run MRP

T Code: MD02

After entering the MD02, enter the finished part number, plant, MRP area, MRP control parameters, Process control parameters, then enter that MRP will be carried out. Then, after running MRP, save here thus, MRP for this month has been generated automatically. The planned order of 1200 has been generated based on the master data we have uploaded, and currently, there is an available inventory of 187. So, we have followed the procedural steps to run MRP for M160 3.73, and planned orders have been generated. We have a planned order of $800 + 400 = 1200$ for the next month. So, in this way, a plan order will be generated for every month. We need to feed customer demand every month, then we need to run the MRP. In this way, planned orders will be generated automatically.

Understand the material requirements, stock availability, and any potential problems, such as shortages or excess inventory, by analyzing the MRP data. To obtain insight into the planning scenario, make use of the numerous MRP reports and transactions accessible in SAP. Follow the MRP's suggestions when carrying out the scheduled procurement or production activities. This involves creating purchase orders or contracts, turning planned orders into production orders, and starting the required manufacturing or procurement processes. Track the progress of procurement and production, keep updated on how planned activities are being carried out, and update the system with actual data when orders are fulfilled. This makes it possible to track inventory levels, production status, and order status accurately. Review and improve the MRP process frequently based on customer input, performance data, and business needs. To increase overall MRP effectiveness, adjust planning parameters, check the accuracy of forecasts, and identify opportunities for process improvement.

4 Results and Discussion

Table 6 shows the planned order results for M160 3.73 and M180 4.88 gears. Table 6 depicts that a planned order of 1200 units for the gear of 160 3.73 for the following month, which was produced as a result of the SAP MRP installation in the gear plant, shows that the system was successful in determining the material requirements and creating a production plan to satisfy the expected demand.

Furthermore, we had 1600 planned orders and 227 available inventory for the MRP for 180 4.88 gear. Based on the creation of a projected order of 1200 units, it seems that the SAP MRP system accurately predicted the demand for gears in the upcoming month.

Table 6 Planned order results for M160 3.73 and M180 4.88 gears

Type of gear	Planned orders for next month	Available inventory
M160 3.73	1200	187
M180 4.88	1600	227

5 Conclusion

Make certain that products are available for customer delivery and that raw materials are available for production. MRP plays a vital role in preparing manufacturing schedules, delivery schedules, and purchase orders to keep inventory levels at a minimum. Many production planning concepts can be assimilated in a realistic environment by combining commercial software with extensive real-world planning. Product costs can be reduced with the appropriate MRP, allowing businesses to produce more goods at cost-effective prices. This chapter studies the implementation of SAP-based MRP in the gear manufacturer organization. As a result of implementing SAP MRP, demand fluctuation can be adjusted, waste can be reduced, shortages can be prevented, cost efficiency can be increased, material requirements can be more accurately estimated, and orders can be placed on the appropriate schedule. This fosters a competitive atmosphere that makes things more affordable, quicker, dependable, and adaptable.

6 Future Scope

In this study, demand uncertainty has not been considered. Uncertainty will also be taken into account in the future to present a more accurate organizational model. In future studies, the application of machine learning can be utilized for better forecasting of customer demand. Some robust MRP tools can also be adopted for better production planning.

References

Angolia M. G., and Pagliari L. R. (2018) "Experiential Learning for Logistics and Supply Chain Management Using an SAP ERP Software Simulation", Decision Sciences Journal of Innovative Education, Volume 16 (2) 104-125, https://doi.org/10.1111/dsji.12146.

Ardiyanto A., Amaanullah D., Fadhillah F. F., Adhipratama R, and Nurcahyo R., (2022) "Material Requirements Planning for PT Petrokimia Gresik's Fertilizer Production to Reach Production Cost Efficiency" Proceedings of the 5th European International Conference on Industrial Engineering and Operations Management Rome, Italy, July 26–28, 2022. available at: https://ieomsociety.org/proceedings/2022rome/247.pdf

Atmika N. D., Tarigan T. M., Annisa Y., and Nurdini A. (2022) "Optimizing Inventory Management in Micro Small Medium Enterprise (MSME) Using Material Requirement Planning (MRP)". Proceedings of the 5th European International Conference on Industrial Engineering and Operations Management Rome, Italy, July 26–28, 2022. available at: https://ieomsociety.org/proceedings/2022rome/242.pdf

Hartoyo P. U. P. and Arvitrida N. I. (2023) "Design and Implementation of MRP-JIT Hybrid System Under Demand Uncertainty" Proceedings of the 10th International Conference on Operations and Supply Chain Management, 2020, Wellington, New Zealand, 14–16 December 2020.

Hasanati N., Permatasari E., Nurhasanah N., and Hidayat S. (2018) Implementation of Material Requirement Planning (MRP) on Raw Material Order Planning System for Garment Industry, IOP Conference Series Materials Science and Engineering 528(1):012064, https://doi.org/10.1088/1757-899X/528/1/012064

Marchello D. Jessica, Gozali L., and Ali A. (2022) "Analysis of Raw Material Planning with Optimal Cost Consideration Using Safety Stocks, Material Requirement Planning, and Lot Sizing Methods at Plastic Jar Company" available at: Proceedings of the 3rd Asia Pacific International Conference on Industrial Engineering and Operations Management, Johor Bahru, Malaysia, September 13–15, 2022 https://www.researchgate.net/publication/367404731

Kasat A. G., and Deshmukh H. M., (2020) "Productivity Improvement through MRP in a manufacturing industry-A", International Research Journal of Engineering and Technology, 7(9) 2361–2364, https://www.irjet.net/archives/V7/i9/IRJET-V7I9421.pdf

Kumar, P., Bhonde, V. N., Rajak, S., & Sundararaman, M. (2025). Machine learning applications in inventory management. *Industry 4.0 for Manufacturing Systems: Concepts, Technologies, and Applications*, 137.

Malindzakova M., Garaj P., Trpcevska J. and Malindzak D. (2022) "Setting MRP Parameters and Optimizing the Production Planning Process". ***Processes*** 10(4), 690; https://doi.org/10.3390/pr10040690

Rajak, S., Kumar, P., Modi, A., Swarnakar, V., Antony, J., & Sony, M. (2025). An assessment of barriers to integrate lean six sigma and industry 4.0 in manufacturing environment: case based approach. *International Journal of Computer Integrated Manufacturing*, *38*(3), 386–407.

Sato R., Matsui Y., and Furukado M. (2016). "An Integrated Planning Mechanism for Supply Chain Management with Routing Subproblem" International journal of Industrial Engineering and Management science. available at: https://web.archive.org/web/20180421102315id. http://www.ijiems.com/files/Concept%20of%20Bills%20of%20Material%20for%20Supply%20Chain%20Planning%20and%20Its%20.pdf

A Structured Approach to Overcome AI Adoption Barriers in the Indian Healthcare Supply Chain: An ISM-Based Analysis

Ambuj Kumar and Om Ji Shukla

Abstract Artificial Intelligence (AI) in the Indian healthcare supply chain (HSC) faces numerous problems, necessitating an organized approach to overcome obstacles and ensure smooth integration. Interpretive Structural Modeling (ISM) offers a systematic framework for analyzing the interdependencies between these barriers and developing strategic solutions. The study identifies high driving power barriers as primary obstacles, including high implementation costs (C1), high initial investment (C9), a lack of skilled labor (C3), and infrastructure integration challenges (C5), and recommends government incentives, targeted training programs, and middleware solutions to overcome them. Regulatory and data-related hurdles (C8, C7) need the creation of AI compliance frameworks, and standardized data protocols to improve system reliability. Awareness campaigns and explainable AI strategies to boost stakeholder confidence must address adoption resistance and trust difficulties (C10, C12). By focusing on fundamental roadblocks stepwise to additionally resolve other related challenges, ISM provides a roadmap for how AI can best be adopted to scale in the India Healthcare value chain while keeping the process efficient, transparent, and scalable.

Keywords Interpretive Structural Modeling (ISM) · AI adoption · Healthcare supply chain · India · Adoption barriers · Regulatory challenges

1 Introduction

AI revolutionizing global healthcare: By 2030, its market will be more than $200 billion (Samineni et al. 2023). In resource-constrained India with limited and inequitable access to healthcare services in a large population, the potential of AI to create impact is immense. All these components are parts of the HSC, which is

A. Kumar (✉) · O. J. Shukla
Department of Mechanical Engineering, National Institute of Technology Patna, Bihar, India
e-mail: ambujk.ph21.me@nitp.ac.in

M. Ram et al. (eds.), *Technological Advancements in Engineering and Management Sciences*, Management and Industrial Engineering,
https://doi.org/10.1007/978-3-032-04606-2_3

involved in the process of medical acquisition, storage, and distribution for the products that are required for the provision of patient care (Beaulieu and Bentahar 2021). AI is expected to make processes more efficient, resources more effectively allocated, and costs more affordable in a multitude of domains. This chapter examines the special challenges that India's healthcare supply chain faces when adopting AI technologies, uncovers technology, legal, economic, and socio-cultural limits, and offers tangible solutions for a more sustainable and equitable future.

AI is revolutionizing healthcare through predictive analytics, robots, natural language processing, and machine learning. (Olawade et al. 2024). These technologies enable systems to process and analyze massive amounts of data, spot trends, and reach good conclusions. For health care, AI will have the most direct impact on the supply chain, leading to added efficiency and accuracy. For instance, AI-driven demand forecasting can aid in predicting medical supply needs, minimizing shortages and waste. The management of stocks is controlled in real time and optimized to guarantee that vital goods are in the right place at the right time. AI improves patient care by using predictive and forecasted analytics to detect at-risk patients and optimize treatment plans (Alowais et al. 2023). AI also plays a role in spotting scams, as it can sift through data on billing and transactions to highlight unusual activity—and, ultimately, prevent victims from losing money. These are some of the applications that are contributing to make the healthcare ecosystem stronger and effective.

The private sector offers modern health care services, but is geographically focused in urban environments, leaving rural areas underserved. Hospital, pharmaceutical companies, government organizations, and logistics providers are essential stakeholders in the Indian HSC. As important as it is to provide medical supply services, VSCs face various challenges, including dispersed infrastructure, inefficient inventory management, and a lack of system integration. These are compounded by geographic diversity, resource limitations, and the high volume of the population, making the delivery of healthcare services fair across the country increasingly problematic. This book chapter discusses the challenges of AI adoption in the HSC system, concentrating on the Indian scenario. The chapter seeks to:

1. Contextualize Challenges for India: Focus on specific challenges posed by India's heterogeneous and complex healthcare landscape, urban–rural gap, scale of the population, and constraints of resources through a comprehensive literature survey.
2. Assessing the challenges: Assessing the difficulties derived from applying the ISM method to determine the critical barriers and enablers.
3. Suggest Recommendations: Advise practical solutions and measures to leap over the hurdles, accelerate the adoption of AI, and enhance healthcare in India.

2 Literature Review

There are many challenges in the adoption of AI, particularly in healthcare and the SC in India. Limited infrastructure, cybersecurity issues, and low government support have been identified as the major barriers to AI adoption in health care (Markopoulou and Papakonstantinou 2021). There are many challenges in the adoption of AI in healthcare and the SC in India. Limited infrastructure, cybersecurity issues, and low government support have been identified as the major barriers to AI adoption in health care (Silveira et al. 2023). At the AI uptake phase in SCRM, various factors play varying roles in the manufacturing, wholesale, retail, and transportation sectors (Hamdi et al. 2018). However, despite these challenges, AI holds the promise of making supply chains more efficient, especially in e-commerce, where it enhances efficiency by predicting demand, by optimizing inventory, and by achieving faster delivery (Modgil et al. 2022). These challenges are to be overcome by fulfilling the infrastructure, addressing the issue of data integration, and worker adaptations toward AI technologies in various industries in India. Table 1 shows the list of recent literature on the adoption barriers to AI applications in healthcare.

Current research on AI-driven systems has numerous significant limitations that limit their effectiveness and uptake. A big challenge is the real-time data integration, i.e., the inability of the AI friend to make fast decisions. In addition, scalability challenges prevent AI models from dealing with large amounts of data and growing demand in a powerful way. Data privacy issues and poor interoperability are cited as the main barriers to AI adoption, as organizations fail to send data in a secure and frustration-free way. The absence of standard frameworks compounds the challenges, as it makes it difficult to establish consistency within the AI systems. In addition, the absence of AI-empowered risk assessment models, as well as the uncertainty in AI model predictions, poses questions on the reliability and accuracy of the decision-making. Regulatory barriers and a lack of data sharing slow AI development, particularly in tightly regulated industries. Additionally, there is currently little work on real-time AI interventions, and a dearth of use cases, making it difficult to evaluate the applied impact of AI solutions. The absence of lifecycle analysis models also restricts the capacity to assess AI performance and sustainability over the longer term. Filling such gaps is vital for the further development of AI systems and for ensuring their effectiveness in a wide range of applications.

There are barriers to adoption of mature technologies (like AI and more advanced technologies. One of the key challenges is the high cost of implementation and initial investment, which many institutions and businesses, especially small-medium enterprises (SMEs), cannot bear for deploying solutions driven by AI. In addition, reluctance to adopt technology and resistance from traditional suppliers hinder efforts to transform digitally. The AI talent shortage is stifling adoption as companies struggle to find the people needed to build, operate, and optimize AI systems. Technical complexity and coexistence with legacy systems impose supplementary challenges, thus requiring significant effort for smooth deployment. Issues of ethics (AI-like decision-making) and trust (such as we don't trust AI suggestions) make the users

Table 1 List of recent literature on AI adoption barrier in the healthcare supply chain

Authors	Major contribution	AI applications in healthcare supply chain	Research gap	Adoption barrier challenges
Badhan et al. (2024)	AI-driven inventory optimization in healthcare supply chains	Predictive analytics for demand forecasting	Lack of real-time data integration	High implementation cost (C1)
Gupta et al. (2022)	Blockchain and AI for pharmaceutical logistics	AI-based counterfeit drug detection	Scalability issues	Resistance to technology adoption (C2)
Boretti (2024)	AI-based predictive maintenance for medical equipment	IoT & AI for monitoring medical device performance	Data privacy concerns	Lack of skilled workforce (C3)
Abolghasemi et al.(2025)	Machine learning for optimizing hospital supply chains	AI for automated supplier selection	Limited interoperability	Ethical concerns in AI decision-making (C4)
Goodarzian et al. (2023)	AI in cold chain management for vaccines	AI-powered temperature monitoring	Limited AI-driven risk assessment models	Integration with existing infrastructure (C5)
Damoah et al. (2021)	Role of AI in reducing medicine shortages	Deep learning for stock prediction	Lack of standardized frameworks	Data security risks (C6)
Revanna and Al-Nakash (2023)	AI-based route optimization for medical deliveries	AI-driven logistics management	Uncertainty in AI model predictions	High dependency on data quality (C7)
Lokanan and Maddhesia (2025)	AI for fraud detection in medical supply procurement	NLP for detecting fraudulent claims	Regulatory challenges	Lack of regulatory clarity (C8)
Shamsuddoha et al. (2025)	AI-enhanced supply chain transparency	AI for supplier performance assessment	Limited research on real-time AI interventions	High initial investment (C9)
Yeo et al. (2022)	AI-powered decision support systems in procurement	AI for optimizing purchase orders	Data-sharing limitations	Resistance from traditional suppliers (C10)

(continued)

Table 1 (continued)

Authors	Major contribution	AI applications in healthcare supply chain	Research gap	Adoption barrier challenges
Han et al. (2023)	AI and digital twins for hospital inventory management	Digital twins for real-time inventory tracking	Insufficient case studies	Technical complexity (C11)
Dewasiri et al. (2024)	AI-driven sustainability in healthcare supply chains	AI for reducing medical waste	Lack of lifecycle analysis models	Trust issues in AI recommendations (C12)

and related stakeholders distrustful. Security threats to data and dependence on high-quality data also add to concerns about the reliability and safety of the AI applications. Its lack of regulatory clarity also means that the rules of the game are fuzzy, which makes it more difficult for companies to navigate through compliance. Confronting these limitations is central to growing AI adoption, ensuring higher levels of AI utility and potential impact across a range of industries.

3 Methodology

The found parameters and ISM model were confirmed by empirical evidence from the experts knowledgeable about healthcare logistics and/or AI. A Delphi approach was employed in which experts reassessed and adjusted the linkages between the components in iterative rounds until consensus was achieved. This helped in keeping the robustness and real-life orientation of the model intact, and thus, the model can be considered as a handy instrument for the policy-makers in their endeavor to improve the AI leveraging in HSC.

This study uses the ISM approach to examine the use of AI in the healthcare supply chain in India, as shown in Fig. 1. This aids in understanding how various elements influence one another and which act as major drivers. The elements are divided into four groups: autonomous, dependent, linkage, and independent. This classification aids in identifying the most important variables that must be prioritized for successful AI implementation.

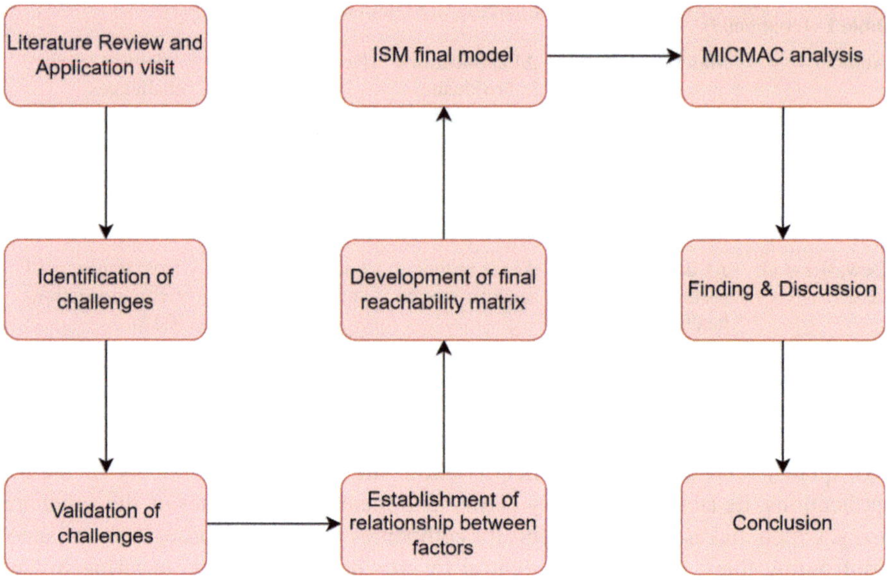

Fig. 1 Methodology adopted in this study

4 ISM Implementation

4.1 SSIM for AI Adoption Barriers in the Indian HSC

The SSIM is developed to establish the contextual relationships among the 12 identified challenges influencing AI adoption in the healthcare supply chain. The relationships between these challenges are determined based on expert opinions and the ISM approach.

4.2 Developing the Initial Reachability Matrix (IRM)

The IRM is derived, as shown in Table 2, from the SSIM by converting the relationships into binary values (0s and 1s) based on the following transformation rules:

Table 2 IRM of the challenges

Challenges	C1	C2	C3	C4	C5	C6	C7	C8	C9	C10	C11	C12
C1	1	1	1	0	1	1	1	1	1	0	1	1
C2	0	1	1	0	0	0	0	0	1	1	1	0
C3	0	0	1	0	0	0	1	0	1	0	1	1
C4	0	0	0	1	0	1	0	1	0	0	1	1
C5	0	0	0	0	1	1	1	1	1	0	1	1
C6	0	0	0	0	0	1	0	1	0	0	1	1
C7	0	0	0	0	0	0	1	1	1	0	1	1
C8	0	0	0	0	0	0	0	1	1	0	1	1
C9	1	0	0	0	0	0	0	0	1	0	1	1
C10	0	0	0	0	0	0	0	1	0	1	1	1
C11	0	0	0	0	0	0	0	0	0	0	1	1
C12	0	0	0	0	0	0	0	0	0	0	0	1

4.3 Developing the Final Reachability Matrix (FRM)

The FRM as shown in Table 3 is obtained by incorporating the transitivity property, ensuring that if C(i) → C(j) and C(j) → C(k), then C(i) → C(k). This is done iteratively.

Table 3 FRM of the challenges

Challenges	C1	C2	C3	C4	C5	C6	C7	C8	C9	C10	C11	C12
C1	1	1	1	0	1	1	1	1	1	0	1	1
C2	0	1	1	0	0	0	0	0	1	1	1	0
C3	0	0	1	0	0	0	1	0	1	0	1	1
C4	0	0	0	1	0	1	0	1	0	0	1	1
C5	0	0	0	0	1	1	1	1	1	0	1	1
C6	0	0	0	0	0	1	0	1	0	0	1	1
C7	0	0	0	0	0	0	1	1	1	0	1	1
C8	0	0	0	0	0	0	0	1	1	0	1	1
C9	1	0	0	0	0	0	0	0	1	0	1	1
C10	0	0	0	0	0	0	0	1	0	1	1	1
C11	0	0	0	0	0	0	0	0	0	0	1	1
C12	0	0	0	0	0	0	0	0	0	0	0	1

Table 4 Level partitioning table for AI adoption barriers

Level	Challenges (Barriers)
Level 6 (Root causes)	C1 (High implementation cost), C2 (Resistance to technology adoption), C3 (Lack of skilled workforce), C4 (Ethical concerns in AI Decision-making), C6 (Data security risks)
Level 5	C5 (Integration with existing infrastructure), C7 (High dependency on data quality)
Level 4	C8 (Lack of regulatory clarity), C9 (High initial investment)
Level 3	C10 (Resistance from traditional suppliers)
Level 2	C11 (Technical complexity)
Level 1 (most dependent)	C12 (Trust Issues in AI recommendations)

Fig. 2 Final ISM model

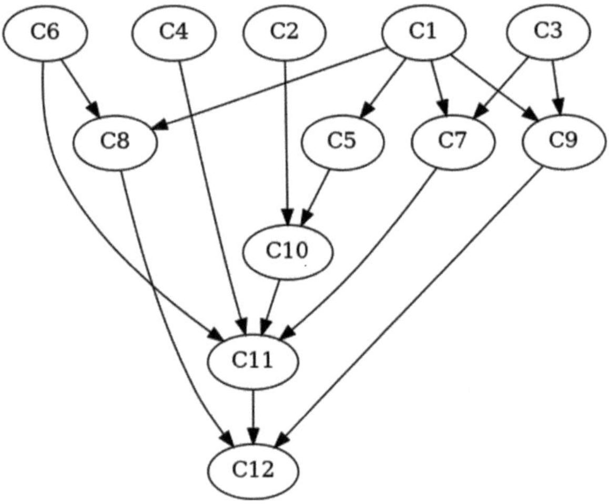

4.4 Level Partitioning

Table 4 and Fig. 2 show the challenges in order of level. This structured table and diagram help in understanding which factors need to be tackled first to resolve AI adoption barriers in India's healthcare supply chain

4.5 MICMAC Analysis

A MICMAC study is used to categorize the observed obstacles based on their Driving Power (impact on other factors) and Dependence Power. Driving Power is calculated by calculating the number of factors that a specific barrier affects (row-wise sum in the FRM). Dependence Power is calculated by calculating the number of factors that influence a specific barrier (column-wise total in the FRM). Table 5 shows the dependence and driving force of challenges.

Based on the Driving and Dependence Power levels, challenges are characterized as follows:

1. Independent Factors: These are the key driving forces that must be addressed to enable AI adoption.
   - C1 (High Implementation Cost)
   - C5 (Integration with Existing Infrastructure)
   - C7 (High Dependency on Data Quality).

   These key driving forces must be addressed first to enable AI adoption.
2. Linkage Factors: These are highly interactive and require strategic interventions as they affect and are affected by multiple other barriers.
   - C8 (Lack of Regulatory Clarity)
   - C9 (High Initial Investment).
3. Dependent Factors: These factors are consequences of other issues and will improve as the independent and linkage factors are addressed.

Table 5 Dependence and driving force of challenges

Factor	Driving force (Rows sum)	Dependence force (column sum)
C1 (High implementation cost)	7	3
C2 (Resistance to technology adoption)	5	4
C3 (Lack of skilled workforce)	6	3
C4 (Ethical concerns in AI decision-making)	5	3
C5 (Integration with existing infrastructure)	7	4
C6 (Data security risks)	5	3
C7 (High dependency on data quality)	6	4
C8 (Lack of regulatory clarity)	6	5
C9 (High initial investment)	6	4
C10 (Resistance from traditional suppliers)	5	4
C11 (Technical complexity)	4	5
C12 (Trust Issues in AI recommendations)	3	6

- C10 (Resistance from Traditional Suppliers)
- C11 (Technical Complexity)
- C12 (Trust Issues in AI Recommendations).

4. Autonomous Factors: These have weak interdependencies and may not significantly impact the system unless triggered by external influences.

- C2 (Resistance to Technology Adoption)
- C3 (Lack of Skilled Workforce)
- C4 (Ethical Concerns in AI Decision-Making)
- C6 (Data Security Risks).

5 Recommendations for AI Adoption in the Indian Healthcare Supply Chain

This section explains major recommendations for AI adoption in the Indian healthcare supply chain, focusing on regulatory support, infrastructure development, AI-powered logistics, fraud detection, and workforce upskilling.

5.1 ISM Recommendations

ISM helps structure the complex interdependencies among barriers and provides a clear roadmap for addressing them.

Focus on Key Driving Barriers (Level 6 & 5):

- **C1 (High Implementation Cost)** and **C9 (High Initial Investment):** Implement government incentives, subsidies, and public–private partnerships to reduce financial constraints.
- **C3 (Lack of Skilled Workforce):** Develop AI-specific training programs and certification courses for healthcare supply chain professionals.
- **C5 (Integration with Existing Infrastructure):** Invest in AI-compatible systems and middleware solutions to facilitate smooth adoption.

Address Regulatory and Data-Related Barriers (Level 4):

- **C8 (Lack of Regulatory Clarity):** Establish AI guidelines and compliance frameworks with government and industry collaboration.
- **C7 (High Dependency on Data Quality):** Implement standardized data collection, storage, and validation protocols.

Tackle Adoption Resistance and Trust Issues (Levels 3, 2, 1):

- **C10 (Resistance from Traditional Suppliers):** Conduct awareness campaigns and showcase AI benefits to reduce hesitation among suppliers.

- **C12 (Trust Issues in AI Recommendations):** Ensure explainable AI (XAI) practices, transparency in AI decisions, and rigorous validation in real-world settings.

5.2 Micmac

It categorizes barriers based on driving and dependence power, helping prioritize interventions.

Key Recommendations are as follows:

1. **Prioritize Independent Factors (High Driving Power, Low Dependence)**
   - Address **C1, C5, and C7** first, as they are root causes that influence multiple challenges.
   - Focus on cost reduction, infrastructure integration, and data quality improvements.
2. **Manage Linkage Factors (High Driving, High Dependence)**
   - **C8 (Regulatory Clarity) and C9 (Investment)** are crucial and interlinked; resolving them will positively impact AI adoption.
3. **Monitor Dependent Factors (Low Driving, High Dependence)**
   - **C10, C11, and C12** will improve once cost, infrastructure, and regulatory issues are resolved.
   - These require trust-building measures, training, and policy changes.
4. **Track Autonomous Factors (Low Driving, Low Dependence)**
   - **C2, C3, C4, and C6** are relatively isolated; they may not directly influence AI adoption, but should still be considered for long-term sustainability.

6 Findings and Discussion

The study's findings highlight the primary challenges to AI adoption in India's HSC, emphasizing the importance of an organized strategy to addressing them. Cost-related concerns, including high costs of adoption and the need for up-front investment, are identified as leading factors in adoption decisions. Targeted strategies involving government subsidies, investment in increments, and public–private partnership should also be introduced to ease the financial strains of healthcare institutions. Additionally, regulatory uncertainty and concerns related to data are significant obstacles, as the absence of established standards, as well as the dependence on high-quality data, represent barriers for the adoption of AI. In order to create an AI future, we can rely on the establishment of standardized regulations, compliance frameworks, and robust data governance processes. When traditional suppliers aren't ready to use AI insights, it can make it harder for companies to adopt them. AI-powered processes are not widely adopted by many stakeholders because they are doubtful about safety and transparency. This issue can be resolved by raising AI awareness, starting small projects, and pointing to successful uses of AI. Clear and transparent explainable AI will help beat doubts about AI. A tiered approach where near-term moves focus on cutting costs and managing rules, and long-term actions put more stress on the fair use of AI, and preparing employees should be adopted. All these actions, along with overcoming the barriers, will depend on partnering among public bodies, business, and education organizations. When these issues are handled logically, the healthcare SC can use AI and become more efficient, transparent, and able to scale up. Incorporating AI into a business can be made quicker if budget cuts, clearer rules, and updated staffing are set as principal aims. Finance and policy development are typical short-term solutions applied here. Still, a business can only succeed in AI long term if it relies on trust and ethical applications of the technology. All stakeholders, whether government, industry, or academia, have to make efforts together to deal with obstacles, fund improvements, and secure skilled employees. With this system, India's healthcare supply chain can make good use of AI, making it more efficient, clearer, and flexible.

7 Conclusion, Future Work, and Limitations

A hierarchical approach is used here to look closely at the most important challenges to adopting AI in the healthcare SC in India. According to the findings, finance, regulation, and crafting the right workforce are the most difficult issues, impacting additional difficulties such as supplier scepticism and trust in the advice provided by AI. By supporting government assistance, suitable regulatory norms, and preparing the workforce, it will become easier to integrate AI. To achieve success with the technology, a method should begin by removing financial and legal obstacles and continuing with actions to make the infrastructure and stakeholders feel confident about it.

This study is useful, though it has a few flaws. The findings come from using an interpretive structural approach that relies on expert opinions and can sometimes be biased. The analysis centers mainly on the Indian healthcare supply chain, so it doesn't apply easily to other sectors or businesses. The future changes in AI and its technology could influence whether some key challenges will still affect software end users. To make the conclusions stronger, future research can conduct case studies in actual industries and get a wider range of input from industry experts. Experts can conduct further research by using both machine learning methods and large-scale industry data. Research also shows that comparing different nations' approaches helps us see in detail how economic and regulatory factors shape AI adoption across the world. Important technologies on the horizon, such as Blockchain for managing data safely and federated learning for privacy-preserving AI applications, ought to be examined. To improve adoption of data in healthcare logistics, future work should analyze the lasting effects of introducing AI, such as reduced expenses, improved efficiencies, and both patient health and safety.

References

Abolghasemi, Mahdi, Babak Abbasi, and Zahra HosseiniFard. 2025. "Machine Learning for Satisficing Operational Decision Making: A Case Study in Blood Supply Chain." International Journal of Forecasting 41 (1): 3–19. https://doi.org/10.1016/j.ijforecast.2023.05.004

Alowais, Shuroug A, Sahar S Alghamdi, Nada Alsuhebany, Tariq Alqahtani, Abdulrahman I Alshaya, Sumaya N Almohareb, Atheer Aldairem, et al. 2023. "Revolutionizing Healthcare: The Role of Artificial Intelligence in Clinical Practice." BMC Medical Education 23 (1): 689. https://doi.org/10.1186/s12909-023-04698-z

Badhan, Istiaque Ahmed, Moohtasim Haque Neeroj, and Irfan Chowdhury. 2024. "The effect of ai-driven inventory management systems on healthcare outcomes and supply chain performance: a data-driven analysis." Frontline Marketing, Management and Economics Journal 4 (11): 15–52. https://doi.org/10.37547/marketing-fmmej-04-11-03

Beaulieu, Martin, and Omar Bentahar. 2021. "Digitalization of the Healthcare Supply Chain: A Roadmap to Generate Benefits and Effectively Support Healthcare Delivery." Technological Forecasting and Social Change 167:120717. https://doi.org/10.1016/j.techfore.2021.120717

Boretti, Alberto. 2024. "A Narrative Review of AI-Driven Predictive Maintenance in Medical 3D Printing." The International Journal of Advanced Manufacturing Technology 134 (5): 3013–24. https://doi.org/10.1007/s00170-024-14305-0

Damoah, Isaac Sakyi, Anthony Ayakwah, and Ishmael Tingbani. 2021. "Artificial Intelligence (AI)-Enhanced Medical Drones in the Healthcare Supply Chain (HSC) for Sustainability Development: A Case Study." Journal of Cleaner Production 328:129598. https://doi.org/10.1016/j.jclepro.2021.129598

Dewasiri, Narayanage Jayantha, Mananage Shanika Hansini Rathnasiri, and Karunarathnage Sajith Senaka Nuwansiri Karunarathna. 2024. "Artificial Intelligence-Driven Technologies for Environmental Sustainability in the Healthcare Industry." In Transforming Healthcare Sector Through Artificial Intelligence and Environmental Sustainability, edited by Rubee Singh, Wasswa Shafik, David Crowther, and Vikas Kumar, 67–87. Singapore: Springer Nature Singapore. https://doi.org/10.1007/978-981-97-9555-0_4

Goodarzian, Fariba, Ali Navaei, Behdad Ehsani, Peiman Ghasemi, and Jesús Muñuzuri. 2023. "Designing an Integrated Responsive-Green-Cold Vaccine Supply Chain Network Using Internet-of-Things: Artificial Intelligence-Based Solutions." Annals of Operations Research 328 (1): 531–75. https://doi.org/10.1007/s10479-022-04713-4

Gupta, Anil Kumar, Gaurang Vivek Awatade, Suyog Sanjay Padole, and Yash Santosh Choudhari. 2022. "Digital Supply Chain Management Using AI, ML and Blockchain." In Innovative Supply Chain Management via Digitalization and Artificial Intelligence, edited by Kumaresan Perumal, Chiranji Lal Chowdhary, and Logan Chella, 1–19. Singapore: Springer Singapore. https://doi.org/10.1007/978-981-19-0240-6_1

Hamdi, Faiza, Ahmed Ghorbel, Faouzi Masmoudi, and Lionel Dupont. 2018. "Optimization of a Supply Portfolio in the Context of Supply Chain Risk Management: Literature Review." Journal of Intelligent Manufacturing 29 (4): 763–88. https://doi.org/10.1007/s10845-015-1128-3

Han, Yilong, Yinbo Li, Yongkui Li, Bin Yang, and Lingyan Cao. 2023. "Digital Twinning for Smart Hospital Operations: Framework and Proof of Concept." Technology in Society 74:102317. https://doi.org/10.1016/j.techsoc.2023.102317

Kumar, Ambuj, Om Ji Shukla, and Shailesh Mani Pandey. 2024. "An Integrated Fuzzy DEMATEL-BWM Method to Analyse Critical Factors for Sustainable Vaccine Supply Chain in Low-Middle Income Nations." International Journal of Multicriteria Decision Making 10 (1): 70–101. https://doi.org/10.1504/IJMCDM.2024.143268

Lokanan, Mark E, and Vikas Maddhesia. 2025. "Supply Chain Fraud Prediction with Machine Learning and Artificial Intelligence." International Journal of Production Research 63 (1): 286–313. https://doi.org/10.1080/00207543.2024.2361434

Markopoulou, Dimitra, and Vagelis Papakonstantinou. 2021. "The Regulatory Framework for the Protection of Critical Infrastructures against Cyberthreats: Identifying Shortcomings and Addressing Future Challenges: The Case of the Health Sector in Particular." Computer Law & Security Review 41:105502. https://doi.org/10.1016/j.clsr.2020.105502

Modgil, Sachin, Rohit Kumar Singh, and Claire Hannibal. 2022. "Artificial Intelligence for Supply Chain Resilience: Learning from Covid-19." The International Journal of Logistics Management 33 (4): 1246–68. https://doi.org/10.1108/IJLM-02-2021-0094

Olawade, David B, Aanuoluwapo C David-Olawade, Ojima Z Wada, Akinsola J Asaolu, Temitope Adereni, and Jonathan Ling. 2024. "Artificial Intelligence in Healthcare Delivery: Prospects and Pitfalls." Journal of Medicine, Surgery, and Public Health 3:100108. https://doi.org/10.1016/j.glmedi.2024.100108

Revanna, Jai Keerthy Chowlur, and Nushwan Yousif B Al-Nakash. 2023. "Metaheuristic Link Prediction (MLP) Using AI Based ACO-GA Optimization Model for Solving Vehicle Routing Problem." International Journal of Information Technology 15 (7): 3425–39. https://doi.org/10.1007/s41870-023-01378-5

Silveira, Franco da, Sabrina Letícia Couto da Silva, Filipe Molinar Machado, Jayme Garcia Arnal Barbedo, and Fernando Gonçalves Amaral. 2023. "Farmers' Perception of the Barriers That Hinder the Implementation of Agriculture 4.0." Agricultural Systems 208:103656. https://doi.org/10.1016/j.agsy.2023.103656

Samineni, Ramu, Guru M prasad, Souvik Giri, Murakonda Likhitha, Yepuri Rasi Anjaneswari, Sinka Padmavathi, M Guru Prasad, and Archana Pattanaik. n.d. "Artificial Intelligence Role in Health Care: Transforming Future Medicine." https://www.researchgate.net/publication/369375959

Shamsuddoha, Mohammad, Eijaz Ahmed Khan, Md Maruf Hossan Chowdhury, and Tasnuba Nasir. 2025. "Revolutionizing Supply Chains: Unleashing the Power of AI-Driven Intelligent Automation and Real-Time Information Flow." Information (Switzerland). Multidisciplinary Digital Publishing Institute (MDPI). https://doi.org/10.3390/info16010026

Yeo, Sook Fern, Cheng Ling Tan, Ajay Kumar, Kim Hua Tan, and Jee Kit Wong. 2022. "Investigating the Impact of AI-Powered Technologies on Instagrammers' Purchase Decisions in Digitalization Era–A Study of the Fashion and Apparel Industry." Technological Forecasting and Social Change 177:121551. https://doi.org/10.1016/j.techfore.2022.121551

Machine Learning for Stock Market Forecasting: Trends and Techniques

Jigyasha Arora and Suyash Bhardwaj

Abstract The expansion of financial markets and the simultaneous increase in complex stock market offenses necessitated the development of innovative detection and prevention measures. This review article offers a comprehensive analysis of advancements in the intersection of stock crime detection and machine learning methodologies for the last thirty years. This research attempts to thoroughly investigate the progression of machine learning applications in the identification of fraudulent activities, insider trading, and market manipulation. This examination is conducted through an in-depth investigation of relevant literature, empirical case studies, and rigorous methodological assessments. The historical trajectory demonstrates the evolution of countermeasures devised in response to affluent individuals engaged in illicit financial activities. The field of machine learning has demonstrated a continuous progression to effectively respond to the intricate dynamics observed within financial markets. The advancement of statistical methodologies and ensemble techniques in subsequent years played a vital part in overcoming the gap between the initial rule-based systems and the transformative capabilities of deep learning models. The paper highlights the significant role that machine learning may play in enhancing the capabilities of market participants and financial regulatory bodies, which enhances the precision of predicting market irregularities and detecting fraudulent conduct by facilitating the analysis of extensive volumes of financial data. This paper addresses ethical concerns about data protection, neutrality, and transparency in utilizing these tools. The provided text serves as an initial foundation for comprehending the historical context, present challenges, and prospective advancements in the realm of preventing stock market crime.

Keywords Financial markets · Stock crime detection · Machine learning · Data privacy

J. Arora (✉) · S. Bhardwaj
Department of Computer Science and Engineering, Faculty of Engineering and Technology, Gurukula Kangri Deemed to be University, Haridwar, India
e-mail: jigyashaarora.cs@gmail.com

S. Bhardwaj
e-mail: suyash.bhardwaj@gkv.ac.in

M. Ram et al. (eds.), *Technological Advancements in Engineering and Management Sciences*, Management and Industrial Engineering,
https://doi.org/10.1007/978-3-032-04606-2_4

1 Introduction

Machine Learning is a compelling emerging technology. The area of machine learning is experiencing rapid growth within the realm of data science. The prediction of financial data poses a significant difficulty (Kumar et al. 2022). Machine learning (ML) is a field of study that is closely intertwined with computer applications, which can enhance their performance through experiential learning. The potential exists for the development of self-customizing systems. The efficacy and efficiency of machine learning techniques in addressing social issues have been significantly enhanced as a result of the availability of extensive datasets and improvements in technology by many organizations. Machine learning and data mining techniques are utilized to identify fraudulent credit card (CC) applications and detect instances of transaction fraud. Bayesian networks, decision trees (DT), support vector machines (SVM), k-nearest neighbor (KNN), neural networks, hidden Markov models (HMM), artificial immune systems (AIS), and self-organizing maps (SOM) are all examples of machine learning algorithms that are often used in many different fields (Li 2019).

A fraud protection device serves to safeguard intricate technological systems by mitigating fraudulent activities. Nevertheless, this does not serve as a deterrent to fraudulent activities. The implementation of fraud detection measures is highly recommended to enhance the security of technical systems. Credit card fraud detection notifies the system administrator about instances of fraudulent transactions. This operation, which is both thrilling and delicate, necessitates accurate identification and detection. Machines may encounter difficulties in accurately identifying instances of credit card fraud. In order to identify instances of fraudulent activity, a system must undergo thorough training using relevant data (Jenga et al. 2023). Machine learning encompasses a range of statistical techniques, such as clustering, regression, and classification. We have selected machine learning-based techniques for fraud detection due to the following rationales:

The system consistently identifies instances of fraudulent activity, operates in real time, expedites the authentication process, and uncovers hidden data associations. Bank loans are essential for the financial landscape. All financial institutions strive to employ efficient commercial strategies to incentivize consumers to engage in borrowing activities (DaCorte 2022). Certain clients exhibit negative behavior following the approval of their application. To effectively cater to consumers, banks must anticipate their behavioral patterns. In this context, financial institutions employ machine learning algorithms, which demonstrate strong performance.

Supervised learning methodologies employ historical transactions recorded by cardholders or credit card companies to predict the occurrence of fraudulent activities in new transactions. To implement this approach, it is necessary to possess a dataset that includes instances of deceitful observations and nondeceitful observations. The utilization of unsupervised learning necessitates the application of clustering techniques to unlabeled data. The proponents of this perspective claim that the presence of irregularities serves as an indication of fraudulent activity. The process of clustering is utilized to identify distinct data distributions that necessitate the implementation of diverse prediction models (DaCorte 2022). Semi-supervised approaches employ unsupervised techniques to detect possible instances of fraudulent transactions and retrieve recently occurring fraudulent transactions. Support Vector Machines (SVM), Multilayer Perceptron (MLP), Random Forest Regression, Autoencoder, and Isolation Forest are potential methods that can be integrated to effectively identify fraudulent credit card transactions. The identification of deceptive practices in e-commerce platforms is presently a major and much-debated topic. To effectively resist fraudulent actions, it is crucial to possess a comprehensive awareness of the diverse approaches and methods deployed by fraudsters. The formation of a robust and sophisticated fraud-detecting mechanism designed to address the challenges posed by rapidly evolving fraudulent activity utilizing the data that CyberSource has provided Based on research published in 2017, an analysis of the magnitude of fraud losses can be conducted by considering the order channel. The online retail platform experienced a significant decline of 74% in fraudulent activities. Based on the findings of a study conducted by Kumar et al. (2022), a notable disparity in the utilization of mobile channels is observed between two distinct groups. Specifically, Group A reports a usage rate of 51 percent, while Group B reports a slightly lower usage rate of 49 percent. Based on the available information, it can be inferred that the primary aim is to detect anomalies within the dataset. The deceptive behavioral inclinations have undergone modifications when examined retrospectively.

An effective fraud detection system should possess the capability to identify instances of rapid fraudulent transactions. The areas of focus in this study are fraud exposure and misuse identification. The process of normalizing abnormalities in transactions is undertaken to acquire knowledge and employ innovative techniques for detecting and preventing fraudulent activities. The misuse of fraud detection systems involves utilizing labeled transaction data, which consists of both regular and fraudulent activities, to train the historical database. The present approach for detecting misuse incorporates both supervised and unsupervised learning techniques to identify anomalies. What are the distinguishing factors or characteristics that set something apart from others? What are the differences between supervised and unsupervised learning?

The labeled datasets used in supervised learning labeled datasets are crucial in the training process and contribute to the enhancement of accuracy. Modifying the settings of the learning rate subsequently, the dataset learning rate parameters are employed through supervised learning techniques, such as the multilayer perceptron (MLP), to construct the model based on historical data stored in a database. Supervised learning is not without limitations, as it may classify new instances of fraud as genuine if they do not align with existing database records. Unsupervised learning is a method that acquires novel transactions and identifies abnormalities and previously unknown transaction patterns. Unsupervised learning necessitates the utilization of methodologies for anomaly detection (Pumsirirat and Yan 2018).

2 Literature Review

2.1 Global Financial Fraud Landscape

Fraud is a prevalent global problem that has a damaging impact on various enterprises, leading to major negative outcomes for both companies and their stakeholders. Based on current estimations, the occurrence of fraudulent activities on a worldwide level over the last two decades has led to significant financial consequences amounting to an astonishing total of $5.127 trillion. Moreover, there has been a notable rise of 56% in the losses associated with these activities during the past decade (Craja et al. 2020). The potential expenses linked to fraudulent activities encompass more than just the immediate cash consequences. It is imperative to consider the indirect expenses associated with a scandal, including the negative impact on investor, creditor, and employee confidence, as well as the consequent decrease in company operations. Over time, fraudulent practices can lead to financial insolvency. The occurrence of fraudulent activities has a widespread and significant influence on various industries and organizations. However, it is worth noting that smaller enterprises, namely those with fewer than 100 workers, as well as non-profit groups, may have less robust internal control framework and limited support to effectively manage and prevent fraud. As a result, these entities may be more susceptible to fraudulent actions.

The responsibility of identifying irregularities in financial statements is considered to be the obligation of the auditor (Cecchini et al. 2010). Despite the established principles, the identification of symptoms of fraudulent activity can present significant challenges. According to a report published in 2020, the identification of fraudulent cases by both internal and external auditors was found to be relatively low, with 15% and 4%, equivalent rates (Albrecht et al. 2008). Consequently, there has been a heightened emphasis on the development of robotic methods to identify instances of deceitful financial statements (Beneish 1997). Many stakeholder groups place a great deal of weight on these systems. Investors benefit from them as they enable informed decision-making. Auditing businesses find value in these systems as they enhance the efficiency and accuracy of audits. State regulators also benefit from these systems as they allow for a more efficient concentration of investigations (Glancy and Yadav 2011). Consequently, significant endeavors have been undertaken to devise intelligent systems specifically engineered in an effort to detect cases of financial statement fraud. Prior research has investigated a range of linguistic and quantitative finance variables as potential guide to economic abnormalities (Beneish 1997). Within the framework of yearly reporting, numerical financial data is accompanied by written material, such as the Management's Discussion and Analysis (MD&A) section (Mehbodniya et al. 2021). The purpose of this textual component is to furnish investors with an understanding of the management's perspectives on the organization's future potential. The linguistic patterns employed within the MD&A section possess the potential to reveal the cognitive mechanisms of managers and serve as indicators of fraudulent conduct. While previous studies have underscored the growing importance of conducting textual analysis on financial paperwork, there is a notable gap in research about the utilization of advanced models for deep learning (DL) for extracting textual features. Moreover, there is a lack of extensive scholarly investigation into the integration of financial and linguistic data to accurately predict instances of financial statement fraud. Additionally, limited attention has been given to the interpretability of these predictions, a critical factor in providing adequate help to auditors.

2.2 *Insider Trading Detection System*

Insider trading is a pervasive type of unethical behavior observed inside the stock market that has persisted since the inception of the securities market. Regulatory bodies across the globe encounter considerable obstacles in efficiently tackling the matter of insider trading. India has demonstrated a notably inadequate record in terms of punishing those implicated in insider trading. Throughout the preceding three decades, the Securities and Exchange Board of India (SEBI) has not successfully obtained any convictions about the act of insider trading. This study's main objective is to examine key indicators in the period preceding the disclosure of insider knowledge to determine the features of insider trading. Throughout our investigation, we have established a methodology to detect occurrences of insider trading within the Indian stock market. In the first step, we acquired corporate filing data from the official website of the National Stock Exchange (NSE) for all the companies enlisted in the NIFTY 50 index. This data specifically pertains to instances of insider trading that took place within the Indian financial market. The dataset encompassed multiple columns that contained data about pricing, action, and the entities or institutions associated with these activities. The duration of data collection encompassed the time frame starting on January 1st and concluding on the day before the release of financial outcomes for the December quarter of the fiscal year 2020–21.

After careful analysis, it has been noted that notable transactions have occurred before the dissemination of financial statements within some firms, giving rise to concerns over possible instances of insider trading. During that particular point in time, machine learning algorithms were utilized with the objective of training and making predictions on occurrences of insider trading. Following this, the algorithms were utilized to train and predict occurrences of insider trading. In the end, the performance of the participants was measured, contrasted, and subsequently graded for accuracy. The experimental results indicated that the suggested methodology successfully achieved the maximum level of precision. The potential benefits of this technique extend beyond the Indian stock market, encompassing various stock exchanges and effectively identifying instances of insider trading. The suggested techniques and conclusions in this study possess considerable importance, as they offer valuable insights for market authorities aiming to improve their efficacy and precision in monitoring insider trading activity (Sapkota et al. 2021).

2.3 Fighting Financial Crime with AI and ML

Artificial intelligence (AI), machine learning (ML), and cybersecurity are broad fields within the realm of computer science distinguished by their notable progress and advancement. Artificial intelligence (AI) encompasses the cognitive talents and capacities demonstrated by machines specifically engineered to acquire knowledge and understanding through human experiences and intellectual processes. These machines possess the capability to adjust to novel inputs and execute tasks that exhibit characteristics similar to those of human behavior. Cybersecurity involves the deployment and exploitation of proactive strategies to safeguard the networks, data, programs, and security systems of an organization from digital threats, unauthorized access, and potential harm. The term "Big Data" encompasses the practice of examining large quantities of data, which can be categorized as structured, semi-structured, or unstructured, and is derived from a wide range of sources that possess complex attributes. Moreover, the effective execution of this activity requires the technology of cutting-edge analytics programs and robust data-processing software. Financial crimes refer to illicit actions involving the improper transfer of property ownership from individuals, corporations, or governments to personally benefit the culprit or perpetrators. Financial crimes cover a broad range of illicit actions that exhibit varying degrees of complexity and scale. These operations are carried out by individuals or small groups, as well as by organized criminal networks, criminal enterprises, and even terrorist organizations. Money laundering and terrorist financing are widely recognized as significant types of financial crimes. A wide range of illegal behaviors, including but not limited to fraud, are included in financial crimes (such as check fraud, credit card fraud, mortgage fraud, medical fraud, corporate fraud, securities fraud, insider trading, bank fraud, insurance fraud, market manipulation, payment fraud, point of sale fraud, and health care fraud), tax evasion, embezzlement, forgery, counterfeiting, bribery, corruption, electric crimes, and identity theft (Mehbodniya et al. 2021).

2.4 Fraud Detection in Credit Card Using Machine Learning

The extensive use of credit cards and the ongoing progression of digital payment systems are increasing day by day. Nevertheless, the task of adequately monitoring instances of credit card theft presents a significant cost challenge for several service providers. Hence, it is crucial to enhance the fraud detection system. The repeated instances of fraudulent scenarios have a substantial effect on financial losses. A variety of technological techniques, such as phishing and virus-infected Trojans, are commonly utilized to obtain confidential information related to credit cards and the personal information of their owners. Therefore, it is crucial to implement effective technology solutions to detect diverse types of fraudulent activity linked to credit card transactions. This study utilizes distinguished machine learning and deep learning approaches to detect occurrences of credit card fraud. In this study, different methods, such as Naive Bayes, Logistic Regression, K-Nearest Neighbor (KNN), Random Forest, and the Sequential Convolutional Neural Network, are used to train on both normal and unusual transaction features. The objective is to identify fraudulent behavior in credit card transactions. The precision of the model is evaluated using publicly accessible data. The accuracy of multiple algorithms, specifically Naive Bayes, Logistic Regression, K-Nearest Neighbor (KNN), Random Forest, and the Sequential Convolutional Neural Network, was represented as 96.1%, 94.8%, 95.89%, 97.58%, and 92.3%, respectively. The relative analysis reveals that the KNN algorithm produces better outcomes in comparison to other methods [18].

3 Results and Discussion

In this review paper, machine learning is discussed about how it has changed stock crime detection. For the past three decades, machine learning has been applied to the investigation and prevention of fraudulent activity, secret trading, and manipulative practices in financial markets.

An investigation into the development of different approaches to machine learning over time revealed that rule-based systems were utilized in the past, whereas deep learning strategies are utilized today. In the 1990s, people started building rule-based algorithms to discover odd trading trends. These algorithms are still used today. These strategies were not successful in keeping up with the ever-evolving marketplace and increasingly intricate criminal operations.

In the 2000s, statistical methods advanced to the point where they were able to look at historical data and discover anomalous patterns. To forecast unusual occurrences on the market over the following ten years, ensemble methods combined the results of several different algorithms. These advancements demonstrated that accuracy could be increased by machine learning and speed of stock crime identification. Techniques for "deep learning," such as RNNs and CNNs, brought about significant shifts in the industry in the 2010s. These neural networks were particularly good at discovering complex temporal and geographical relationships in the financial data, which made it easier to forecast anomalous behavior. This was one of the benefits of using these neural networks. Because there was a large amount of data and a significant amount of computer power at their disposal, deep learning models were able to handle complex financial data.

4 Conclusion

In summarizing its findings, this review study illustrates how machine learning has had a significant influence on stock crime identification and how that identification has evolved over the past three decades. The fact that the discipline has progressed from straightforward rule-based systems to complex deep learning algorithms demonstrates how dedicated it is to combat the ever-evolving nature of financial crimes. Not only has the use of machine learning approaches made detection more precise and effective, but it has also demonstrated how crucial it is for administrative agencies, financial professionals, and data scientists to participate.

Because the world of finance is in a constant state of flux, it is of the utmost importance that machine learning and the investigation of stock crimes work together. It is necessary to manage ethical problems such as data privacy, fairness, and transparency with attention to ensure that these tools are used appropriately. The capacity of machine learning to adjust and come up with new ideas will be an essential component of being able to successfully cope with the ways that financial criminals are adopting new techniques in the future. This ability will be necessary to prepare for future developments. In forthcoming research, new approaches to problem-solving should be investigated, with due consideration given to the rapid pace of technological advancement and the ever-changing financial crime scene.

In the end, the purpose of this review article is to assist individuals in gaining a more comprehensive grasp of the relationship between machine learning and the identification of unlawful stock trading actions. This article takes a comprehensive look at historical achievements, ongoing challenges, and potential achievements in the foreseeable future. It will do this by making it less difficult to learn, by encouraging individuals to collaborate on projects, and by strengthening the global financial system to lower the likelihood of illegal activity in the stock market.

References

Albrecht, W. S., Albrecht, C., & Albrecht, C. C. (2008). Current Trends in Fraud and its Detection. *Information Security Journal A Global Perspective*, *17*(1), 2–12. https://doi.org/10.1080/19393550801934331

Beneish, M. D. (1997). Detecting GAAP violation: implications for assessing earnings management among firms with extreme financial performance. *Journal of Accounting and Public Policy*, *16*(3), 271–309. https://doi.org/10.1016/s0278-4254(97)00023-9

Cecchini, M., Aytug, H., Koehler, G. J., & Pathak, P. (2010). Making words work: Using financial text as a predictor of financial events. *Decision Support Systems*, *50*(1), 164–175. https://doi.org/10.1016/j.dss.2010.07.012

Craja, P., Kim, A., & Lessmann, S. (2020). Deep learning for detecting financial statement fraud. *Decision Support Systems*, *139*(113421), 113421. https://doi.org/10.1016/j.dss.2020.113421

DaCorte, A.M. *The effects of the internet on financial institutions' fraud mitigation.* (2022).

Glancy, F. H., & Yadav, S. B. (2011). A computational model for financial reporting fraud detection. *Decision Support Systems*, *50*(3), 595–601. https://doi.org/10.1016/j.dss.2010.08.010

Huang, S.-Y., Tsaih, R.-H., & Yu, F. (2014). Topological pattern discovery and feature extraction for fraudulent financial reporting. *Expert Systems with Applications*, *41*(9), 4360–4372. https://doi.org/10.1016/j.eswa.2014.01.012

Jenga, K., Catal, C., & Kar, G. (2023). Machine learning in crime prediction. Journal of Ambient Intelligence and Humanized Computing, 14(3), 2887-2913.

Kumar, S., Ahmed, R., Bharany, S., Shuaib, M., Ahmad, T., Tag Eldin, E., Rehman, A. U., & Shafiq, M. (2022). Exploitation of machine learning algorithms for detecting financial crimes based on customers' behavior. *Sustainability*, *14*(21), 13875. https://doi.org/10.3390/su142113875.

Li, Y. (2019). Credit risk prediction based on machine learning methods. *2019 14th International Conference on Computer Science & Education (ICCSE).*

Mehbodniya, A., Alam, I., Pande, S., Neware, R., Rane, K. P., Shabaz, M., & Madhavan, M. V. (2021). Financial fraud detection in healthcare using machine learning and deep learning techniques. *Security and Communication Networks*, *2021*, 1–8. https://doi.org/10.1155/2021/9293877

Öğüt, H., Aktaş, R., Alp, A., & Doğanay, M. M. (2009). Prediction of financial information manipulation by using support vector machine and probabilistic neural network. *Expert Systems with Applications*, *36*(3), 5419–5423. https://doi.org/10.1016/j.eswa.2008.06.055

Pumsirirat, A., & Yan, L. (2018). Credit Card Fraud Detection using Deep Learning based on Auto-Encoder and Restricted Boltzmann Machine. *International Journal of Advanced Computer Science and Applications: IJACSA*, *9*(1). https://doi.org/10.14569/ijacsa.2018.090103

Haitham Nobanee, Mohamed Elhoseny, Xiaohui Yuan, Noura Metawa (2021) Researchgate.net. Retrieved August 11, 2023, from https://www.researchgate.net/publication/348265248_Fighting_Financial_Crime_with_Artificial_Intelligence_Machine_Learning_Cybersecurity_and_Big_Data

Rezaee, Z. (2005). Causes, consequences, and deterence of financial statement fraud. *Critical Perspectives on Accounting*, *16*(3), 277–298. https://doi.org/10.1016/s1045-2354(03)00072-8

Sapkota, A., Kumar, A., & Mathur, A. (2021). Intelligent system for the detection of insider trading in Indian stock market. *Proceedings of the First International Conference on Combinatorial and Optimization, ICCAP 2021, December 7–8 2021, Chennai, India.*

Seyedhossein, L., & Hashemi, M. R. (2010). Mining information from credit card time series for timelier fraud detection. *2010 5th International Symposium on Telecommunications*

Innovations in Vacuum Pressure Impregnation (VPI) Treatment of Bamboo Fiber for Enhanced Performance in Fiber Reinforced Polymer (FRP) Laminated Composites: Structural, Mechanical, and Thermal Advancements

Yogendra Kumar Verma, Amit Kumar Singh, Sandeep Kumar Singh, M. K. Paswan, and S. Prashanth

Abstract This study explores the impact of Vacuum Pressure Impregnation (VPI), alkali, tannic acid, and phosphoric acid treatments on the mechanical, thermal, and tribological properties of bamboo fiber-based composites. Thermogravimetric analysis (TGA) reveals that phosphoric acid-treated fibers exhibit the highest thermal stability due to phosphorus-based cross-linking, while NaOH-treated fibers show the lowest resistance. X-ray diffraction (XRD) and Fourier-transform infrared (FTIR) spectroscopy highlight structural and chemical modifications, with NaOH enhancing crystallinity, tannic acid improving surface bonding, and phosphoric acid introducing flame-retardant properties. Tensile strength analysis demonstrates peak performance at 1 wt% fiber content, with phosphoric acid-treated fibers achieving the highest strength (290.45 ± 2.85 MPa). At 1.5 wt%, strength declines due to fiber agglomeration. These findings underscore the potential of bamboo fiber composites for structural, automotive, and aerospace applications, reinforcing their role as eco-friendly, high-performance materials.

Keywords Bamboo fiber · VPI · Composite · FTIR · TGA

Y. K. Verma · A. K. Singh (✉) · S. K. Singh · M. K. Paswan · S. Prashanth
Department of Mechanical Engineering, National Institute of Technology Nagaland, Dimapur, India
e-mail: amit.kumar965@yahoo.com

M. Ram et al. (eds.), *Technological Advancements in Engineering and Management Sciences*, Management and Industrial Engineering,
https://doi.org/10.1007/978-3-032-04606-2_5

1 Introduction

The increasing global focus on sustainability has driven extensive research into natural fiber-reinforced composites. Bamboo, with its exceptional mechanical properties and rapid renewability, has emerged as a promising reinforcement for FRP composites. Traditional synthetic fibers such as glass and carbon, while offering superior strength, contribute to environmental concerns due to their energy-intensive production and non-biodegradability. Bamboo fiber-based FRP composites present an eco-friendly alternative with comparable mechanical performance. The increasing demand for sustainable and high-performance materials in engineering and management sciences has driven extensive research into fiber-reinforced polymer (FRP) composites. Among the various natural fibers explored, bamboo fibers have emerged as a promising reinforcement material due to their exceptional mechanical properties, high availability, and eco-friendliness (Shah et al. 2016a; Verma et al. 2023; Kumar et al. 2024). The global shift toward sustainable materials has necessitated the development of bio-based composites that balance performance, durability, and environmental impact. Bamboo fiber-reinforced polymer (BFRP) composites have garnered significant attention due to their lightweight nature, high specific strength, and potential for reducing carbon footprints in various industrial applications (Rassiah et al. 2013). Bamboo, as a natural fiber source, possesses a unique hierarchical structure that provides high tensile strength, durability, and resilience. With a cellulose content of approximately 40–50%, bamboo fibers exhibit superior mechanical properties compared to conventional synthetic fibers such as glass and carbon fibers (Shah et al. 2016b). The advancement in processing techniques, such as vacuum pressure impregnation (VPI), alkaline treatment, and hybridization with nanomaterials, has further enhanced the performance of BFRP laminated composites. These advancements contribute to improved fiber-matrix adhesion, enhanced thermal stability, and superior mechanical properties, making them viable alternatives to traditional FRP composites in aerospace, automotive, and structural applications (Trujillo et al. 2014). The evolution of BFRP composites is rooted in the broader development of natural fiber-reinforced composites (NFRCs). Historically, synthetic fiber-based composites dominated engineering applications due to their superior mechanical and thermal performance. However, environmental concerns, high production costs, and limited biodegradability have necessitated the transition toward sustainable alternatives (Fu et al. 2012). In response, natural fibers such as jute, hemp, flax, and bamboo have been extensively studied for their potential in reinforcing polymer matrices (Laftah 2016). Among these, bamboo has been identified as a highly renewable and fast-growing fiber source, capable of competing with synthetic counterparts in load-bearing applications (Kim et al. 2013). Significant advancements in the development of BFRP composites include chemical modifications such as alkaline treatment, silane functionalization, and enzymatic treatment to enhance fiber-matrix bonding. These treatments improve fiber wettability, remove non-cellulosic constituents such as lignin and hemicellulose, and increase composite durability under varying environmental conditions (Cao et al. 2021). Furthermore, hybridization with nanomaterials such as titanium dioxide (TiO_2) and nanoclay has introduced additional functionalities, including antimicrobial properties, enhanced flame resistance, and improved moisture barrier properties (Lima et al. 2022; Singh and Singh 2023; Singh et al. 2024). The fabrication of high-performance BFRP laminated composites involves various processing techniques aimed at optimizing fiber dispersion, interfacial adhesion, and structural integrity. Traditional processing methods such as hand lay-up, compression molding, and vacuum-assisted resin transfer molding (VARTM) have been widely employed in the production of FRP composites (Ramachandran et al. 2016; Singh et al. 2023). However, recent advancements have introduced novel techniques such as 3D printing, microwave curing, and resin infusion under vacuum pressure to enhance fiber impregnation and reduce void content in the final composite structure (Tavares et al. 2020; Prashanth et al. 2024). One of the most promising advancements in BFRP composite fabrication is the vacuum pressure impregnation (VPI) technique. This method ensures uniform resin penetration into bamboo fibers, leading to enhanced mechanical performance and reduced material defects (Jiang et al. 2014; Verma et al. 2024a, 2025). Additionally, the integration of nanofillers, such as graphene oxide and carbon nanotubes (CNTs), into the polymer matrix has resulted in enhanced stiffness, toughness, and electrical conductivity, opening avenues for multifunctional composite applications (Ho et al. 2012). The mechanical performance of BFRP composites is influenced by factors such as fiber content, orientation, and surface treatment. Studies have demonstrated that VPI-treated bamboo fiber composites exhibit superior tensile and flexural strength compared to untreated counterparts. For instance, tensile strength improvements of up to 30% have been reported upon alkaline treatment of bamboo fibers, attributed to enhanced fiber-matrix adhesion and reduced micro voids (Herrero et al. 2010). Thermogravimetric analysis (TGA) and differential scanning calorimetry (DSC) have been extensively utilized to evaluate the thermal stability of BFRP composites. The inclusion of nanoclay and other thermal stabilizers has been shown to increase degradation temperatures, making these composites suitable for high-temperature applications such as automotive and aerospace structures (Visakh et al. 2012). Furthermore, dynamic mechanical analysis (DMA) has revealed that BFRP composites exhibit high storage modulus and improved damping behavior, indicating their potential in vibration-damping applications (Liu et al. 2012). BFRP laminated composites have found diverse applications across multiple engineering sectors, including automotive, aerospace, construction, and marine industries. In the automotive industry, the need for lightweight and high-strength materials has positioned BFRP composites as viable alternatives to conventional glass fiber-reinforced composites, contributing to fuel efficiency and reduced emissions (Pil et al. 2016). In the construction sector, these composites have been employed in load-bearing structures, bridge reinforcements, and earthquake-resistant building materials due to their high specific strength and environmental sustainability (Yu et al. 2014). From a management science perspective, the adoption of BFRP composites aligns with circular economy principles and sustainable supply chain management. Companies investing in green materials benefit from enhanced corporate social responsibility (CSR) initiatives, reduced environmental impact, and compliance with stringent regulatory standards (Sukmawan et al. 2016). The integration of life cycle assessment (LCA) methodologies has further demonstrated the economic and ecological advantages of BFRP composites, fostering their widespread adoption in global markets (Faruk et al. 2014).

This study aims to explore the technological advancements in engineering and management sciences through the development of innovative bamboo fiber-based FRP laminated composites. The objective is to enhance the mechanical, thermal, and tribological properties of these composites by incorporating vacuum pressure impregnation (VPI) and various chemical treatments, including NaOH, tannic acid, and phosphoric acid. The novelty lies in the integration of optimized bamboo fiber treatment techniques with advanced composite fabrication methods to achieve superior tensile strength, and thermal stability. This research contributes to sustainable material development by leveraging bamboo's eco-friendly nature for high-performance applications in structural and automotive industries.

2 Bamboo Fiber Processing and Treatment Methods

2.1 Vacuum Pressure Impregnation (VPI) Treatment of Bamboo

Vacuum Pressure Impregnation (VPI) is a treatment process (Fig. 1) used to enhance the properties of materials, particularly bamboo, by introducing a preservative or functionalizing agent into the bamboo structure. This process involves subjecting the bamboo to a vacuum, which removes air from the cell lumens, followed by applying pressure to force the impregnation solution into the bamboo's vascular system. Key parameters influencing VPI treatment include vacuum pressure, typically 0.4 MPa, and the impregnation pressure, which is 1.0 MPa. The duration of the treatment, typically 2 h, also plays a crucial role in determining the depth of impregnation. Additionally, the type of impregnation solution, such as preservatives, polymers, or functionalizing agents, and the temperature conditions, 80 °C, influence the effectiveness of the VPI process. This treatment improves bamboo's mechanical strength, durability, and resistance to environmental factors by enhancing its chemical and morphological properties.

2.2 Mechano-Chemical Fiber Extraction Technique

The mechano-chemical fiber extraction technique is a hybrid approach that integrates mechanical and chemical processes to efficiently extract high-quality fibers from bamboo. This method enhances fiber yield, improves mechanical properties, and ensures minimal degradation of the natural fiber structure. The process begins

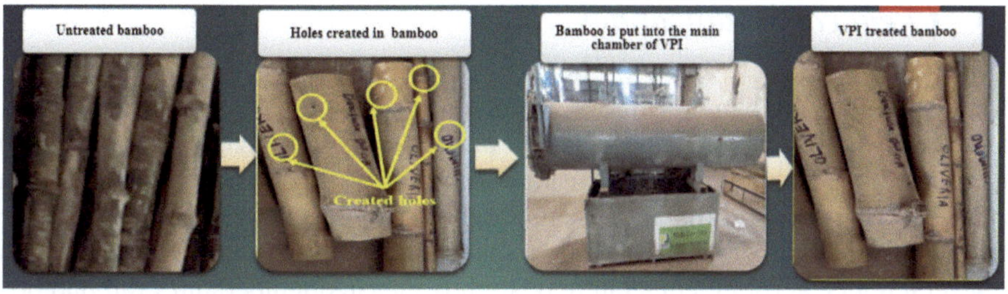

Fig. 1 VPI treatment method of bamboo

with the selection and preparation of raw bamboo culms, which are cut into strips of uniform dimensions. These strips undergo a mechanical breakdown process using a high-speed rotary cutter or milling machine to separate the fibers while maintaining their structural integrity. To further facilitate fiber extraction, the mechanically treated bamboo strips are subjected to a chemical pretreatment process. This step typically involves immersion in an alkaline solution, such as sodium hydroxide (NaOH), at optimized concentrations and treatment durations to remove hemicellulose, lignin, and other non-cellulosic components that bind the fibers together. After the chemical treatment, the fibers are thoroughly washed with distilled water to neutralize any residual chemicals and subsequently dried at controlled temperatures to prevent thermal degradation which are depicted in Fig. 2. The dried fibers are then subjected to additional mechanical processing, such as high-shear refining or ball milling, to further individualize and refine the fibers. The final step involves fiber characterization, including tensile strength testing, and chemical composition evaluation (FTIR and XRD), ensuring that the extracted fibers meet the required structural and performance standards for various applications. This mechano-chemical approach ensures a high fiber yield with improved mechanical properties, making it suitable for advanced composite applications, including bio-composites and nanocellulose production.

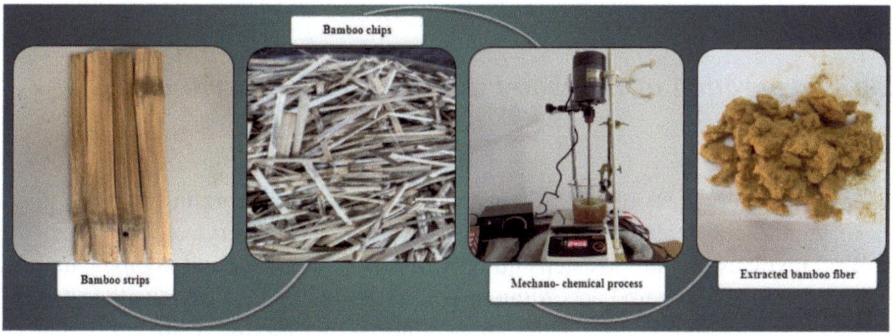

Fig. 2 Mechano-Chemical fiber extraction technique

2.3 Chemical Modifications

Alkali treatment of bamboo fiber, commonly using sodium hydroxide (NaOH), involves several steps to enhance fiber properties by removing impurities and increasing fiber-matrix adhesion. The process begins by immersing bamboo fibers in an NaOH solution, typically at concentrations ranging from 10%, depending on the desired effect. The fibers are soaked for a specified duration, usually between 30 min to several hours, at room temperature or slightly elevated temperatures (around 60–80 °C). This treatment helps to remove hemicellulose, lignin, and other non-cellulosic components, leading to a cleaner, more crystalline cellulose structure. After soaking, the fibers are thoroughly rinsed with distilled water to remove any residual NaOH and neutralized with an acidic solution (like acetic acid) to restore the pH balance. Finally, the fibers are dried at ambient temperature or under controlled conditions to prevent any moisture-related degradation, resulting in enhanced mechanical properties, improved fiber-matrix bonding, and greater compatibility for composite applications. Bamboo fibers were treated with tannic acid (TA) by preparing a 5 mM TA solution, which involved dissolving 0.851 g of TA in 100 ml of distilled water while stirring at 400 rpm until fully dissolved, A tris buffer solution was then added at a 10 mM concentration, and the pH of the resulting solution was adjusted to 8.5 using HCl and NaOH. The bamboo fibers were immersed in this solution for 30 min, followed by thorough rinsing and drying in an oven at 70 °C for 5 h, and then subjected to overnight drying at 105 °C. In the preparation of Phosphorylated Bamboo fibers (P-BFs), the process begins by dissolving 5 g of adenosine diphosphate (ADP) and 10 g of urea in distilled water, stirring the solution at 400 rpm for 15 min, followed by the preparation of a solution comprising ADP, urea, and distilled water, to which 10 g of this solution are added dropwise to a Bamboo fibers (BFs) solution while stirring for 30 min at 500 rpm; the wet BFs obtained are then subjected to drying in a vacuum oven for 24 h at 70 °C, after which they are cured for 2 h at 160 °C, and the phosphorylated fibers are subsequently cooled and thoroughly cleaned with distilled water to eliminate any residual chemicals, culminating in the formation of Phosphorylated Bamboo fibers (P-BFs).

2.4 Preparation of Bamboo Fiber-Based FRP Laminated Composites

The production process of FRP laminated composites incorporating bamboo fibers (BFs), tannic acid-treated bamboo fibers (TABFs), and phosphorylated bamboo fibers (P-BFs) involves meticulous steps utilizing the conventional hand layup technique, with test samples prepared according to ASTM 3039 standard for tensile tests. The process begins with precisely weighing and thoroughly mixing resin and BFs using a mechanical high-shear mixer at 2000 rpm for 20 min, followed by an additional 30 min of stirring at the same speed after introducing the hardener and catalyst to ensure homogeneity. To eliminate potential air bubbles, the viscous slurry undergoes vacuum treatment in a desiccator before being applied in the mold, which is constructed from a sturdy steel flat plate and meticulously prepared with a releasing agent for easy removal. A peel ply is first laid down, followed by the careful arrangement of glass fiber as a layer, after which the slurry mixture, comprising epoxy resin, hardener, accelerator, and filler, is evenly spread across the fibers using a roller, a process that is repeated for up to 10 layers with an additional peel ply added to ensure effortless removal of the laminate from the mold. The mold is then subjected to pre-curing in an oven at 120 °C for 2 h, followed by post-curing at 160 °C for an extended 8 h, ensuring optimal structural integrity and enhanced properties of the final composite product.

3 Structural, Mechanical, and Thermal Characterization

3.1 Thermal Stability and Degradation Analysis of Treated Bamboo Fiber

The thermal stability and degradation behavior of NaOH-, tannic acid-, and phosphoric acid-treated bamboo fibers were analyzed using thermogravimetric analysis (TGA) to evaluate their weight loss characteristics and thermal resistance, which are depicted in Fig. 3. The TGA curves revealed distinct degradation stages corresponding to moisture evaporation, hemicellulose, cellulose, and lignin decomposition. The initial weight loss below 150 °C was attributed to the evaporation of absorbed moisture, with NaOH-treated fibers exhibiting slightly lower moisture loss due to enhanced hydrophobicity. The primary degradation stage, occurring between 250–350 °C, was associated with the decomposition of hemicellulose and cellulose, with phosphoric acid-treated fibers demonstrating the highest thermal resistance, followed by tannic acid- and NaOH-treated fibers (Zhang et al. 2015; Lourenço et al. 2019). The onset temperature of degradation (Tonset) was highest for phosphoric acid-treated fibers, indicating enhanced char formation and improved thermal stability due to phosphorus-based cross-linking. The tannic acid-treated fibers exhibited intermediate stability, attributed to the formation of thermally stable tannin-lignin complexes. NaOH-treated fibers displayed the lowest thermal resistance, with a significant weight loss in the 250–350 °C range, likely due to the partial removal of lignin and the reduction of cellulose crystallinity. The residual char yield at 600 °C followed a similar trend, with phosphoric acid-treated fibers retaining the highest char content, signifying superior thermal stability, while NaOH-treated fibers had the least residual mass. The differences in degradation behavior highlight the role of chemical treatment in modifying the thermal properties of bamboo fibers, with phosphoric acid treatment enhancing thermal stability through cross-linked structures, tannic acid improving resistance via phenolic interactions, and NaOH reducing thermal stability due to structural modifications.

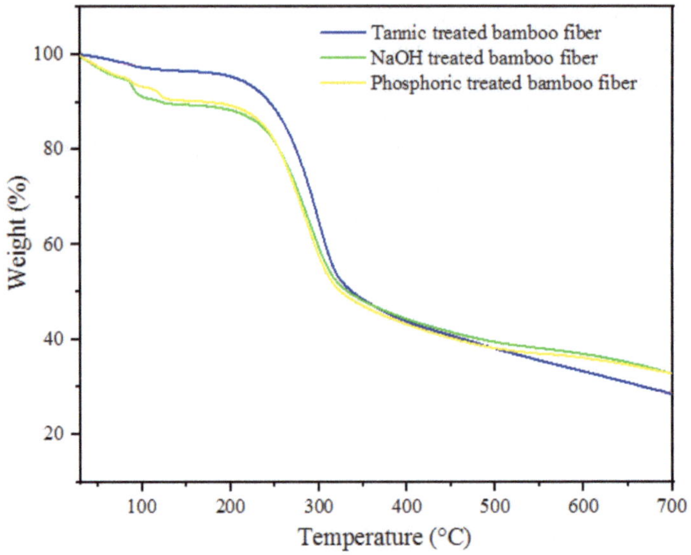

Fig. 3 TGA analysis of treated bamboo fiber

3.2 Characterization (XRD, FTIR) of Treated Bamboo Fiber

The characterization of NaOH-, tannic acid-, and phosphoric acid-treated bamboo fibers was performed using X-ray diffraction (XRD) and Fourier-transform infrared (FTIR) spectroscopy to analyze structural and chemical modifications which are depicted in Fig. 4. XRD analysis revealed that NaOH treatment enhanced the crystalline index of bamboo fibers by effectively removing amorphous components such as hemicellulose and lignin, leading to improved cellulose alignment. In contrast, tannic acid treatment showed moderate crystallinity retention, as the phenolic groups in tannic acid formed hydrogen bonds with the cellulose matrix, preventing excessive degradation of amorphous phases. Phosphoric acid treatment resulted in a lower crystallinity index due to partial hydrolysis and disruption of cellulose chains, leading to structural loosening. FTIR spectra provided insights into functional group modifications; NaOH-treated fibers exhibited increased intensity of O–H stretching at ~ 3400 cm^{-1} due to enhanced cellulose exposure, along with reduced C = O stretching (~1720 cm^{-1}) indicating hemicellulose removal. Tannic acid-treated fibers showed strong absorption peaks at ~ 1600 cm^{-1} and ~ 1500 cm^{-1}, attributed to aromatic C = C stretching from tannin molecules chemically bonding with the fiber surface. Phosphoric acid treatment introduced characteristic P–O stretching vibrations around 1100 cm^{-1}, confirming phosphate ester formation, which may contribute to enhanced flame retardancy. The combined XRD and FTIR analyses demonstrate that NaOH treatment primarily enhances cellulose purity and crystallinity, tannic acid modifies

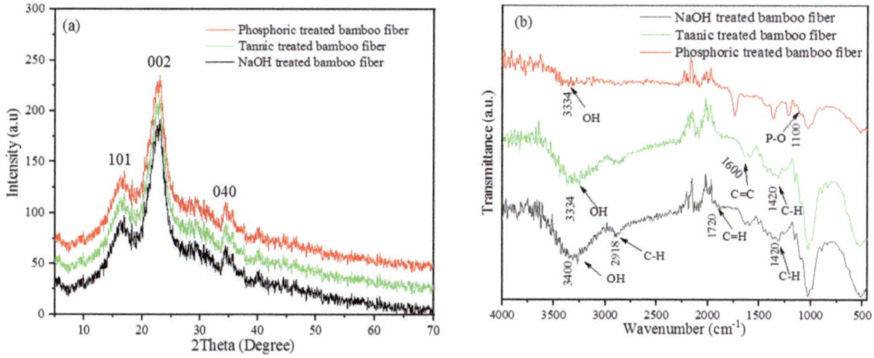

Fig. 4 **a** XRD **b** FTIR analysis of bamboo fiber

the fiber surface through hydrogen bonding interactions, and phosphoric acid introduces chemical modifications that alter both structural and thermal properties (Verma et al. 2024b).

3.3 Tensile Strength of FRP Composite

The tensile strength of bamboo fibers treated with NaOH, tannic acid, and phosphoric acid varies with fiber content (0.5 to 1.5 wt%). At 1 wt%, all treatments exhibit their highest tensile strength values, with phosphoric acid-treated fibers showing the greatest improvement at 290.45 ± 2.85 MPa, followed by tannic acid-treated fibers at 266.78 ± 3.12 MPa and NaOH-treated fibers at 224.56 ± 2.96 MPa. Compared to 0.5 wt%, tensile strength increases for all treatments, indicating enhanced fiber-matrix adhesion and improved interfacial bonding. However, at 1.5 wt%, a decline in tensile strength is observed for all treatments, with NaOH-treated fibers dropping to 210.34 ± 3.14 MPa, tannic acid-treated fibers to 245.89 ± 2.98 MPa, and phosphoric acid-treated fibers to 270.34 ± 3.21 MPa, which are depicted in Fig. 5. This reduction suggests that excessive fiber loading lead to fiber agglomeration, non-uniform stress distribution, and weaker mechanical performance. The superior tensile strength at 1 wt% is attributed to optimal fiber dispersion and improved load transfer within the composite structure, while phosphoric acid treatment yields the highest strength due to enhanced fiber surface modification and better chemical interaction with the matrix.

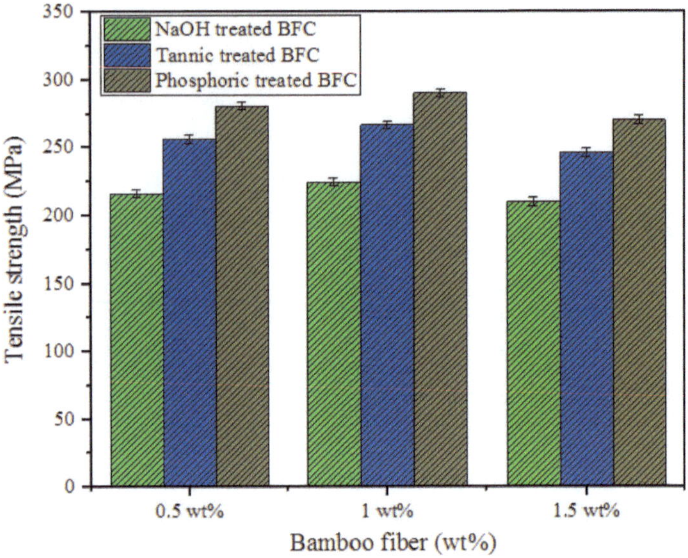

Fig. 5 Tensile Strength of FRP composite

4 Applications and Future Prospects

Recent technological advancements in engineering and management sciences have led to significant innovations in the development of bamboo fiber-based Fiber Reinforced Polymer (FRP) laminated composites. These composites have garnered attention due to their eco-friendly nature, excellent mechanical properties, and sustainability. The integration of bamboo fibers with FRP has enhanced the composites' strength, durability, and resistance to environmental degradation, making them ideal for applications in construction, automotive, and aerospace industries. Future prospects involve further improvements in treatment processes like Vacuum Pressure Impregnation (VPI) and the exploration of hybrid composites with nanoparticles to further enhance the performance of bamboo fiber-based FRP laminated composites. These advancements are expected to drive the widespread adoption of bamboo-based composites, contributing to sustainable material solutions in various industries.

5 Conclusion

The study highlights the impact of chemical treatments on the mechanical, thermal, and structural properties of bamboo fiber-based FRP laminated composites. Phosphoric acid treatment demonstrated superior thermal stability, enhanced char formation, and the highest tensile strength at 1 wt% due to cross-linked structures improving fiber-matrix interactions. Tannic acid-treated fibers exhibited moderate thermal and mechanical performance, attributed to tannin-lignin complex formation, while NaOH treatment enhanced crystallinity but resulted in reduced thermal stability. TGA analysis confirmed distinct degradation patterns, with phosphoric acid-treated fibers retaining the highest residual char. XRD and FTIR analyses provided insights into structural modifications, reinforcing the role of chemical treatments in enhancing composite properties. The tensile strength trend indicated optimal fiber loading at 1 wt%, beyond which agglomeration led to mechanical deterioration. These findings underscore the potential of chemically modified bamboo fibers in high-performance FRP composites for structural and industrial applications.

References

S. Cao, P. Rathi, X. Wu, D. Ghim, Y. S. Jun, and S. Singamaneni, "Cellulose Nanomaterials in Interfacial Evaporators for Desalination: A 'Natural' Choice," *Adv. Mater.*, vol. 33, no. 28, pp. 1–22, 2021, https://doi.org/10.1002/adma.202000922.

T. E. S. de Lima, A. R. G. de Azevedo, M. T. Marvila, V. S. Candido, R. Fediuk, and S. N. Monteiro, "Potential of Using Amazon Natural Fibers to Reinforce Cementitious Composites: A Review," *Polymers (Basel).*, vol. 14, no. 3, pp. 1–20, 2022, https://doi.org/10.3390/polym14030647.

O. Faruk, A. K. Bledzki, H. P. Fink, and M. Sain, "Progress report on natural fiber reinforced composites," *Macromol. Mater. Eng.*, vol. 299, no. 1, pp. 9–26, 2014, https://doi.org/10.1002/mame.201300008.

J. Fu, X. Li, W. Gao, H. Wang, A. Cavaco-Paulo, and C. Silva, "Bio-processing of bamboo fibres for textile applications: A mini review," *Biocatal. Biotransformation*, vol. 30, no. 1, pp. 141–153, 2012, https://doi.org/10.3109/10242422.2012.650450.

M. Herrero, J. A. Mendiola, A. Cifuentes, and E. Ibáñez, "Supercritical fluid extraction: Recent advances and applications," *J. Chromatogr. A*, vol. 1217, no. 16, pp. 2495–2511, 2010, https://doi.org/10.1016/j.chroma.2009.12.019.

M. P. Ho, H. Wang, and K. T. Lau, "Effect of degumming time on silkworm silk fibre for biodegradable polymer composites," *Appl. Surf. Sci.*, vol. 258, no. 8, pp. 3948–3955, 2012, https://doi.org/10.1016/j.apsusc.2011.12.068.

J. Jiang *et al.*, "Evolution of disposable bamboo chopsticks into uniform carbon fibers: A smart strategy to fabricate sustainable anodes for Li-ion batteries," *Energy Environ. Sci.*, vol. 7, no. 8, pp. 2670–2679, 2014, https://doi.org/10.1039/c4ee00602j.

H. Kim, K. Okubo, T. Fujii, and K. Takemura, "Influence of fiber extraction and surface modification on mechanical properties of green composites with bamboo fiber," *J. Adhes. Sci. Technol.*, vol. 27, no. 12, pp. 1348–1358, 2013, https://doi.org/10.1080/01694243.2012.697363.

Y. Kumar, A. Kumar, M. K. Paswan, and P. Kumar, "Preparation and characterization of bamboo based nanocellulose by ball milling and used as a filler for preparation of nanocomposite," *Polymer (Guildf).*, vol. 308, no. July, p. 127396, 2024, https://doi.org/10.1016/j.polymer.2024.127396.

W. A. Laftah and W. A. Wan Abdul Rahman, "Pulping Process and the Potential of Using Non-Wood Pineapple Leaves Fiber for Pulp and Paper Production: A Review," *J. Nat. Fibers*, vol. 13, no. 1, pp. 85–102, 2016, https://doi.org/10.1080/15440478.2014.984060.

K. Liu, H. Takagi, R. Osugi, and Z. Yang, "Effect of physicochemical structure of natural fiber on transverse thermal conductivity of unidirectional abaca/bamboo fiber composites," *Compos. Part A Appl. Sci. Manuf.*, vol. 43, no. 8, pp. 1234–1241, 2012, https://doi.org/10.1016/j.compositesa.2012.02.020.

S. C. Lourenço, M. Moldão-Martins, and V. D. Alves, "Antioxidants of natural plant origins: From sources to food industry applications," *Molecules*, vol. 24, no. 22, pp. 14–16, 2019, https://doi.org/10.3390/molecules24224132.

L. Pil, F. Bensadoun, J. Pariset, and I. Verpoest, "Why are designers fascinated by flax and hemp fibre composites?," *Compos. Part A Appl. Sci. Manuf.*, vol. 83, pp. 193–205, 2016, https://doi.org/10.1016/j.compositesa.2015.11.004.

S. Prashanth, R. Pongen, and Y. K. Verma, "Mechanical and tribological performance of NaOH-Treated bamboo fiber-reinforced recycled HDPE composites under fretting wear conditions," *Wood Mater. Sci. Eng.*, pp. 1–15, 2024, https://doi.org/10.1080/17480272.2024.2399796.

M. Ramachandran, R. Bhargava, and P. P. Raichurkar, "Effect of Nanotechnology in Enhancing Mechanical Properties of Composite Materials," *Int. J. Text. Eng. Process.*, vol. 2, no. 1, pp. 2395–3578, 2016.

K. Rassiah, M. H. Megat Ahmad, M. Kem Sg Besi, and K. Lumpur, "A Review On Mechanical Properties Of Bamboo Fiber Reinforced Polymer Composite," *Aust. J. Basic Appl. Sci.*, vol. 7, no. 8, pp. 247–253, 2013.

A. U. M. Shah, M. T. H. Sultan, M. Jawaid, F. Cardona, and A. R. A. Talib, "A Review on the Tensile Properties of Bamboo Fiber Reinforced Polymer Composites," *BioResources*, vol. 11, no. 4, pp. 10654–10676, 2016a, https://doi.org/10.15376/BIORES.11.4.10654-10676.

A. U. M. Shah, M. T. H. Sultan, M. Jawaid, F. Cardona, and A. R. A. Talib, "A review on the tensile properties of bamboo fiber reinforced polymer composites," *BioResources*, vol. 11, no. 4, pp. 10654–10676, 2016b, https://doi.org/10.15376/biores.11.4.Shah.

S. K. Singh and T. J. Singh, "Study of Mechanical Properties and Water Absorption Behavior of TiO 2 Nanofiller-Enhanced Glass Fiber-Reinforced Polymer Composites : A Review," pp. 799–809.

S. K. Singh, B. Nayak, and T. J. Singh, "Investigating the role of synthesized reduced graphene oxide and graphite micro- fi llers on mechanical and fretting wear performance of glass fi ber epoxy-based composite," vol. 0, no. 0, pp. 1–17, 2023, https://doi.org/10.1177/09540083231196087.

S. K. Singh, T. J. Singh, and B. Nayak, "Analysis of the impact of exfoliated graphene oxide on the mechanical performance and in-plane fracture resistance of epoxy-based nanocomposite," vol. 0, no. 0, pp. 1–21, 2024, https://doi.org/10.1177/09540083241278209.

R. Sukmawan, H. Takagi, and A. N. Nakagaito, "Strength evaluation of cross-ply green composite laminates reinforced by bamboo fiber," *Compos. Part B Eng.*, vol. 84, pp. 9–16, 2016, https://doi.org/10.1016/j.compositesb.2015.08.072.

T. D. Tavares, J. C. Antunes, F. Ferreira, and H. P. Felgueiras, "Biofunctionalization of natural fiber-reinforced biocomposites for biomedical applications," *Biomolecules*, vol. 10, no. 1, 2020, https://doi.org/10.3390/biom10010148.

E. Trujillo, M. Moesen, L. Osorio, A. W. Van Vuure, J. Ivens, and I. Verpoest, "Bamboo fibres for reinforcement in composite materials: Strength Weibull analysis," *Compos. Part A Appl. Sci. Manuf.*, vol. 61, pp. 115–125, 2014, https://doi.org/10.1016/j.compositesa.2014.02.003.

Y. K. Verma, A. K. Singh, M. K. Paswan, and P. K. Gurmaita, "A novel approach to extract and characterize vacuum-pressurized bamboo fibers using mechano-chemical techniques," *J. Wood Chem. Technol.*, vol. 43, no. 4, pp. 1–13, 2023, https://doi.org/10.1080/02773813.2023.2224301.

Y. K. Verma, A. K. Singh, M. K. Paswan, and S. Prashanth, "Mechano-chemical extraction and characterization of tannic acid-functionalized bamboo fibers from vacuum-pressure impregnation (VPI) treated and untreated sources to enhance mechanical properties of fiber-reinforced polymer (FRP) laminated composites," *Wood Mater. Sci. Eng.*, pp. 1–16, 2024, https://doi.org/10.1080/17480272.2024.2436517.

Y. K. Verma, A. K. Singh, M. K. Paswan, and P. Vinay, "Mechanical and tribological characterization of vacuum pressurized impregnation (VPI) treated bamboo fiber-reinforced polymer (FRP) composites with jute mat layer under fretting wear conditions," *Wood Mater. Sci. Eng.*, pp. 1–13, 2024, https://doi.org/10.1080/17480272.2024.2386692.

Y. K. Verma, A. K. Singh, S. K. Singh, S. Dutta, and M. K. Paswan, "Comprehensive analysis of enhanced thermal and mechanical properties in vacuum pressure impregnated (VPI) treated

Chimono bamboo fibers through surface treatment with sodium hydroxide and graphene oxide : in-depth characterization and insights into the performance of graphene oxide-based natural hybrid composites," *J. Wood Chem. Technol.*, vol. 0, no. 0, pp. 1–20, 2025, https://doi.org/10.1080/02773813.2025.2455558

P. M. Visakh, S. Thomas, K. Oksman, and A. P. Mathew, "Crosslinked natural rubber nanocomposites reinforced with cellulose whiskers isolated from bamboo waste: Processing and mechanical/thermal properties," *Compos. Part A Appl. Sci. Manuf.*, vol. 43, no. 4, pp. 735–741, 2012, https://doi.org/10.1016/j.compositesa.2011.12.015.

Y. Yu, X. Huang, and W. Yu, "A novel process to improve yield and mechanical performance of bamboo fiber reinforced composite via mechanical treatments," *Compos. Part B Eng.*, vol. 56, pp. 48–53, 2014, https://doi.org/10.1016/j.compositesb.2013.08.007.

X. Zhang, F. Wang, and L. M. Keer, "Influence of surface modification on the microstructure and thermo-mechanical properties of bamboo fibers," *Materials (Basel).*, vol. 8, no. 10, pp. 6597–6608, 2015, https://doi.org/10.3390/ma8105327.

Machine Learning-Driven Optimization of Waste Heat Recovery in Cement Plants

Sourabh Kumar Singh, Sourabh Luthra, and Anuj Kumar Shukla

Abstract This chapter emphasizes the integration of Machine Learning techniques to enhance waste heat utilization in cement plants. By analyzing operational data, machine learning models can predict and optimize process parameters, enabling better control of heat recovery and improving overall efficiency. These techniques can identify optimal operating conditions to maximize power generation from waste heat and enhance clinker production, providing a data-driven approach to sustainable manufacturing.

Keywords Machine Learning · Waste heat recovery power plants · Cement production

1 Introduction

In modern cement manufacturing, energy efficiency and sustainability are critical challenges, given the high energy demands and environmental impacts of the production process (Kumar and Rakshit 2021). Waste Heat Recovery Power Plants (WHRPPs) have emerged as a transformative solution to address these issues by utilizing the substantial waste heat generated during clinker production in rotary kilns, preheaters, and clinker coolers (Blackburn et al. 2022a; Khater et al. 2021).

WHRPPs convert this unused heat into electricity, meeting 25–30% of a cement plant's energy needs while reducing reliance on external power sources, production costs, and carbon footprints (Amiri and Vaseghi 2015; Gupta and Kaul n.d.). However, the efficiency of WHRPPs depends on several factors, including the number of preheater stages, operational parameters, and design considerations.

This Chapter compares plant performance data with process optimization techniques, analyzing design versus operating parameters and identifying optimal conditions to maximize power generation and clinker production, and also explores the

S. K. Singh · S. Luthra · A. K. Shukla (✉)
Mechanical Engineering Department, NIT Raipur, Raipur, India
e-mail: akshukla.me@nitrr.ac.in

M. Ram et al. (eds.), *Technological Advancements in Engineering and Management Sciences*, Management and Industrial Engineering,
https://doi.org/10.1007/978-3-032-04606-2_6

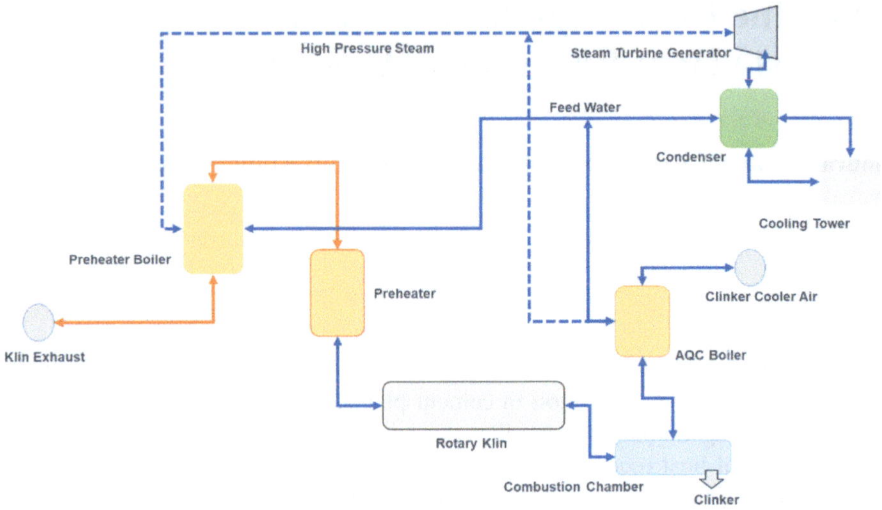

Fig. 1 Schematic diagram of WHRPP in cement plant

integration of Machine Learning techniques to optimize key operating parameters, enabling data-driven insights for maximizing WHRPP efficiency and overall plant performance. By evaluating plant parameters, predicting stress-free operating conditions, and achieving energy consumption targets, the work contributes to advancing sustainable practices and energy efficiency in cement manufacturing.

1.1 Working Cycle

WHRPP installed in the industry, works on Rankine Cycle, the flue gases from the preheater tower and clinker cooler are passed through the boiler to generate steam as shown in Fig. 1 The steam rotates the turbine for power generation (Khalid et al. 2021). The steam from turbine exhaust is cooled through air- or water-cooled condenser, and it is recirculated back to the boiler (Asghar et al. 2023; Blackburn et al. 2022b).

1.2 Brief Introduction About Problem

In the context of waste heat recovery power plants in the cement industry, there is a significant challenge related to boiler design conditions (Dutta et al. 2023).

The abrupt fluctuations in heat input to the boiler, stemming from the variable conditions of the cement manufacturing process, give rise to thermal stress within the boiler (Sivakumar and Holla, n.d.).

This stress causes the damages in boiler tubes which leads to plant outages and a subsequent loss in power generation (Marenco-Porto et al. 2023).

WHRPP operates as varying conditions and at different load and ambient conditions thus the objective of this chapter is to find out the best possible parameters at which we will get maximum output as clinker generation from kiln and also maximize the power generation from the waste heat recovery power plant (Madlool et al. 2012), as shown in Table 1.

1.3 Key Performance Operating Parameters Comparison

From the above parameters it has been observed that under different scenario and combination of mills WHRPP operates above operating parameters hence the same will results unutilized energy due to restriction in boiler and turbine operating conditions or overheating due to high boiler flue gas inlet temperature will create thermal stress in the boiler.

1.4 Benefits of Optimizing Waste Heat Recovery Power Plant

In India, the cement industry holds significant potential for recovering and utilizing waste heat for power generation. As an energy-intensive sector, capturing and reusing waste heat presents an excellent opportunity to lower overall operating costs. With the rise in global greenhouse gas (GHG) emissions, industrial sectors are increasingly pressured to enhance efficiency, optimize operations, and adopt environmentally friendly practices. Leveraging waste heat recovery not only aligns with these goals but also provides a reliable source of power (Fierro et al. 2020; Irungu et al. 2017; Sani et al. 2020).

In cement plants, approximately 35% of heat is lost mainly from preheater and clinker cooler waste gases cement plant can offset approximately 30% of their total electricity requirement. Waste heat recovery power plant offer significant energy saving and contribute to sustainable practices in cement industry (Pranavagam and Watvisave n.d.; Sultanov et al. 2022).

Table 1 Plant key performance parameters comparison

Kiln - Clinker Production (TPD)	Gross Gen/Ton Clinker (KWh)	Average load (MWh)	Gross Gen (KWh)	FG Inlet Temp-PH (°C)	FG Outlet Temp-PH (°C)	Delta Temp. PH (°C)	DP Across PH (MMWC)	FG Inlet Temp AQC (°C)	FG Outlet Temp AQC (°C)	Delta Temp. AQC (°C)	DP Across AQC (MMWC)	Conditions
4500	42.97	8.06	193,344	410.54	176.75	233.79	−53.87	401.22	71.48	329.74	−26.08	Maximum per ton clinker
3345	42.28	5.89	141,432	414.54	169.04	245.5	−49.32	354.37	68.57	285.8	−8.14	Minimum per ton clinker
5385	41.94	9.41	225,863	395.24	176.19	219.05	−53.26	385.23	89.66	295.57	−77.77	Maximum Gross Generation
3345	42.28	5.89	141,432	414.54	169.04	245.5	−49.32	354.37	68.57	285.8	−8.14	Maximum delta T across AQC
4500	42.97	8.06	193,344	410.54	176.75	233.79	−53.87	401.22	71.48	329.74	−26.08	Maximum delta T across PH
3345	42.28	5.89	141,432	414.54	169.04	245.5	−49.32	354.37	68.57	285.8	−8.14	Maximum flue gas inlet Temp PH
6150	31.98	8.19	196,664	324.76	173.96	150.80	−62.32	415.87	95.87	320.00	−73.34	Maximum flue gas inlet Temp AQC

2 Methodology

2.1 Machine Learning Techniques to Optimize Operational Parameters

In Machine Learning, regression models are used to predict continuous outcomes based on input features. Independent parameters, or features are the variables that we use to make these predictions. Here is a general outline of how you can build a regression model with independent parameters as features (Nongthombam and Sharma n.d.).

2.1.1 Data Preprocessing

This step involves cleaning the data, handling missing values, and encoding categorical variables if necessary.

2.1.2 Data Visualization

Choose relevant features that you believe will have a significant impact on the outcome. Feature engineering may involve transforming variables, creating new features, or combining existing ones to better represent the relationships in the data.

2.1.3 Model Evaluation

Evaluate the performance of your model using the testing data. Common evaluation metrics for regression models include Mean Squared Error (MSE), Mean Absolute Error (MAE), root Mean Squared Error (RMSE), and R-squared (Tyagi et al. 2022).

2.1.4 Model Selection

Choose a regression procedure that makes sense for your situation. Among the popular options are support vector regression, polynomial regression, decision tree. Throughout this process, it is important to iterate and refine your model as needed, based on its performance and the specific requirements of your problem (Huang et al., 2020).

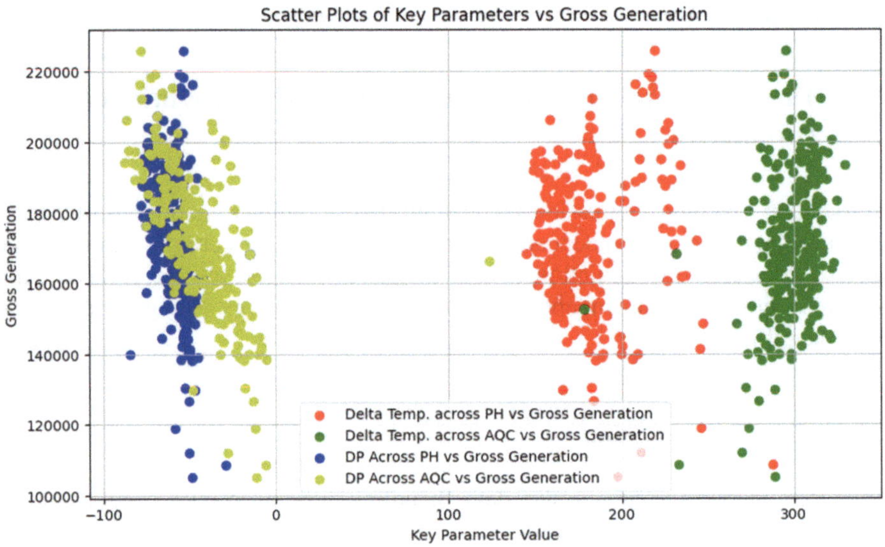

Fig. 2 Scatter plots of key parameters versus Gross generation

2.2 *Data Visualization*

The code will generate a single figure containing scatter plots for each of the key parameters from Table 1.1 (`'DT_PH'`, `'DT_AQC'`, `'DP_PH'`, and `'DP_AQC'`) against `'Gross_gen'`. Each scatter plot will be represented by a different color, and the legend will help distinguish between them (Luthra 2024) (Fig. 2).

2.3 *Plot Key Parameters Versus Clinker Production*

To create the scatter plots for the key parameters versus clinker production, Similar script is used (Fig. 3).

2.4 *To Calculate Z-Scores for Each Column*

A z-score represents both the relative position of an individual score (X) in a distribution as compared to the mean and the variation of scores in the distribution. Standardized scores provide you with a score that is directly comparable within and between different groups of cases (Crewson 2006; Tyagi et al. 2022).

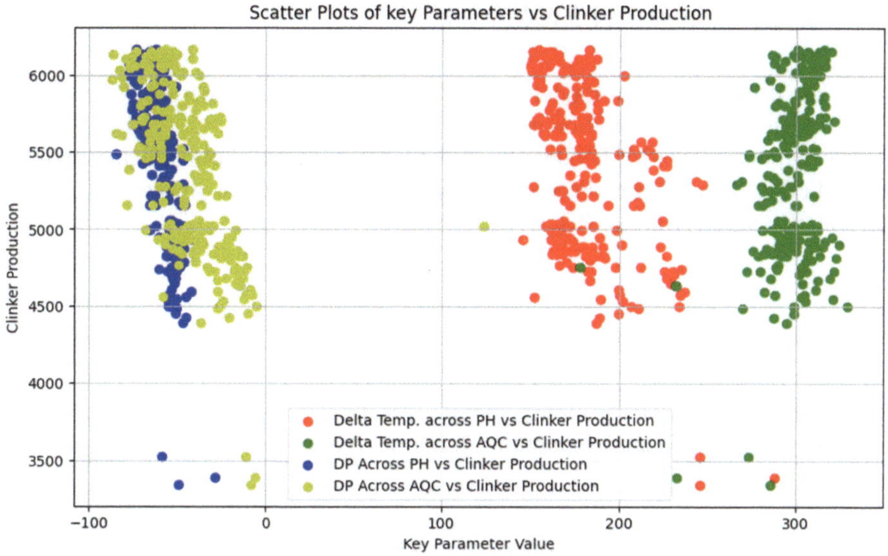

Fig. 3 Scatter plots of key parameters versus Clinker Production

$$Z = \frac{X - \overline{X}}{S}$$

3 Model Evaluation

The code evaluates Machine Learning models for predicting two different outcomes: Clinker Production and Gross Generation.

3.1 Clinker Production Evaluation

The code loops through different Machine Learning models trained to predict Clinker Production. For each model, it calculates Mean Squared Error (MSE) by comparing the predicted values with the actual Clinker Production values.

3.2 Gross Generation Evaluation

Similarly, it loops through models trained to predict Gross Generation. Calculates MSE for each Gross Generation model.

The output provides MSE values for each model, indicating how well they perform in predicting both Clinker Production and Gross Generation, lower MSE values indicate better performance (Crewson 2006).

3.3 Model Selection

Best model for Gross Generation: Gradient Boosting.

Best model for Clinker Production: Random Forest.

4 Optimization Using Grid Search

This code segment performs model selection and parameter optimization for maximizing both Gross Generation and Clinker Production.

4.1 Model Training and Evaluation

It trains models for both Gross Generation and Clinker Production using the provided training data (X_train, y_clinker_train, y_gross_train). Then it evaluates the models on the test set (X_test, y_clinker_test, y_gross_test) using R-squared score (r2_score) and stores the scores(Nongthombam & Sharma, n.d.).

4.2 Identifying Best Models

It identifies the best model for Gross Generation and Clinker Production based on the highest R-squared score.

4.3 Parameter Optimization

It creates a grid of parameter combinations using the scaled parameter space, Predicts Gross Generation and Clinker Production for each parameter combination using the best models, Stores the results including parameter values, Gross Generation, and Clinker Production predictions.

4.4 *Normalization and Combined Score Calculation*

Normalizes Gross Generation and Clinker Production predictions and calculates the harmonic mean of the normalized scores as a combined score for each parameter combination.

4.5 *Identifying Best Parameter Combination*

Finds the parameter combination that maximizes the Gross Generation, Clinker Production, and their harmonic mean.

4.6 *Output*

It prints out the best parameter combinations for maximizing Gross Generation, maximizing Clinker Production, and maximizing both using the harmonic mean.

5 Result and Discussion

From the plant operating set of data applied in ML, the key parameters achieved for maximum clinker production and maximum gross generation are:

Operating Condition	Clinker production (TPD)	Gross generation per ton clinker (KWh/Ton)	Gross generation (KWh)	Delta temp. PH (°C)	DP across PH (°C)	Delta temp. AQC (°C)	DP across AQC (°C)
Max gross generation	5385	41.94	225,863	219.05	−53.26	295.57	−77.77
Max clinker production	6168	28.85	177,952	182.71	−73.63	316.04	−41.55

From Machine Learning Algorithm the Result for Maximum Gross Generation and Maximum Clinker production are as follows:

Best parameters for maximum Gross Generation (actual values):

{'Dt_ph': 200.00001608, 'Dp_ph': −66.66665585999999, 'Dt_aqc': 310.00002268, 'Dp_aqc': −80.00002338}

Best parameters for maximum Clinker Production (actual values):

{'Dt_ph': 182.22224276, 'Dp_ph': −66.66665585999999, 'Dt_aqc': 303.33332719, 'Dp_aqc': −80.00002338}

Best parameters for maximizing both Gross Generation and Clinker Production (actual values):

{'Dt_ph': 200.00001608, 'Dp_ph': −66.66665585999999, 'Dt_aqc': 310.00002268, 'Dp_aqc': −80.00002338}

The actual values after reverse scaling are exactly the same for both maximizing Gross Generation and maximizing the combined score (using Harmonic Mean), it indicates that the parameter values that yield the highest Gross Generation also result in a high Clinker Production, making them the best optimal choice for both objectives. This suggests a significant correlation or interdependence between the parameters influencing both outputs.

6 Conclusion

The primary objective of this chapter is to optimize key operational parameters of the kiln system to enhance both clinker production and gross generation productivity. We focused on key parameters such as the temperature difference in the Preheater (ΔT_PH), pressure drop in the Preheater (DP_PH), temperature difference in the Air Quenching Cooler (ΔT_AQC), and pressure drop in the Air Quenching Cooler (DP_AQC).

The analysis began with data cleaning, which involved identifying and removing outliers to ensure robust model performance. Correlation analysis revealed that ΔT_PH and DP_PH had significant impacts on clinker production, while ΔT_AQC and DP_AQC were critical for gross generation.

The employed advanced Machine Learning techniques, specifically Random Forest and Gradient Boosting, to build the predictive models, these models were trained as per the historical operational data and demonstrated high accuracy in predicting both clinker production and gross generation.

The predictive insights derived from these models allowed us to pinpoint optimal values for the operational parameters, offering a path toward enhanced control and efficiency in kiln operations. The results showed a clear pathway for achieving increased production and energy output by fine-tuning the operational settings based on the optimized parameters, demonstrating the practical utility of our approach in real-world kiln operations.

References

Amiri, A., & Vaseghi, M. R. (2015). Waste heat recovery power generation systems for cement production process. *IEEE Transactions on Industry Applications*, *51*(1), 13–19. https://doi.org/10.1109/TIA.2014.2347196

Asghar, A., Abdul Hussain Ratlamwala, T., Kamal, K., Alkahtani, M., Mohammad, E., & Mathavan, S. (2023). Sustainable operations of a combined cycle power plant using artificial intelligence based power prediction. *Heliyon*, *9*(9). https://doi.org/10.1016/j.heliyon.2023.e19562

Blackburn, L. D., Tuttle, J. F., Andersson, K., Fry, A., & Powell, K. M. (2022a). Development of novel dynamic machine learning-based optimization of a coal-fired power plant. *Computers and Chemical Engineering*, *163*. https://doi.org/10.1016/j.compchemeng.2022.107848

Crewson, P. (2006). *Applied Statistics Handbook*. http://www.acastat.com

Dutta, A., Datta, D., Malebary, S. J., Alam, M. M., Gorji, M. R., & Eldin, S. M. (2023). Development of thermodynamically assisted machine learning model to select best fuel for the thermal power station. *Case Studies in Thermal Engineering*, *49*. https://doi.org/10.1016/j.csite.2023.103208

Fierro, J. J., Escudero-Atehortua, A., Nieto-Londoño, C., Giraldo, M., Jouhara, H., & Wrobel, L. C. (2020). Evaluation of waste heat recovery technologies for the cement industry. *International Journal of Thermofluids*, *7–8*. https://doi.org/10.1016/j.ijft.2020.100040

Gupta, S. K., & Kaul, / S K. (n.d.). *WASTE HEAT RECOVERY POWER PLANTS IN CEMENT INDUSTRY*.

Hedman, B. (2017). *IMPROVING THERMAL AND ELECTRIC ENERGY EFFICIENCY AT CEMENT PLANTS: INTERNATIONAL BEST PRACTICE*. https://doi.org/10.13140/RG.2.2.34495.97447

Huang, J. C., Ko, K. M., Shu, M. H., & Hsu, B. M. (2020). Application and comparison of several machine learning algorithms and their integration models in regression problems. *Neural Computing and Applications*, *32*(10), 5461–5469. https://doi.org/10.1007/s00521-019-04644-5

Irungu, S. N., Muchiri, P., & Byiringiro, J. B. (2017). The generation of power from a cement kiln waste gases: a case study of a plant in Kenya. *Energy Science and Engineering*, *5*(2), 90–99. https://doi.org/10.1002/ese3.153

Khalid, S., Hwang, H., & Kim, H. S. (2021). Real-world data-driven machine-learning-based optimal sensor selection approach for equipment fault detection in a thermal power plant. *Mathematics*, *9*(21). https://doi.org/10.3390/math9212814

Khater, A. M., Soliman, A., Ahmed, T. S., & Ismail, I. M. (2021). Power generation in white cement plants from waste heat recovery using steam-organic combined Rankine cycle. *Case Studies in Chemical and Environmental Engineering*, *4*. https://doi.org/10.1016/j.cscee.2021.100138

Kumar, A., & Rakshit, D. (2021). A critical review on waste heat recovery utilization with special focus on Organic Rankine Cycle applications. In *Cleaner Engineering and Technology* (Vol. 5). Elsevier Ltd. https://doi.org/10.1016/j.clet.2021.100292

Luthra, S. (2024, December 20). *WHRCP_ML_Codes*. https://github.com/luthragithub/whrcp/commit/c5c793c2b44bec3c13385206cf5f6046d6e2d515

Madlool, N. A., Saidur, R., & Rahim, N. A. (2012). Investigation of Waste Heat Recovery in Cement Industry: A Case Study. *International Journal of Engineering and Technology*, *4*(5), 665–667. https://doi.org/10.7763/ijet.2012.v4.457

Marenco-Porto, C. A., Fierro, J. J., Nieto-Londoño, C., Lopera, L., Escudero-Atehortua, A., Giraldo, M., & Jouhara, H. (2023). Potential savings in the cement industry using waste heat recovery technologies. *Energy*, *279*. https://doi.org/10.1016/j.energy.2023.127810

Nongthombam, K., & Sharma, D. (n.d.). *Data Analysis using Python*. www.ijert.org

Pranavagam, D. S., & Watvisave. (n.d.). Design and analysis of waste heat recovery heat exchanger boiler. In *International Engineering Research Journal Special Edition PGCON*.

Sani, M. M., Noorpoor, A., & Motlagh, M. S. (2020). Multi objective optimization of waste heat recovery in cement industry (a case study). *Journal of Thermal Engineering*, *6*(4), 604–618. https://doi.org/10.18186/thermal.764536

Sivakumar, T., & Holla, R. (n.d.). *Design and Simulation of a Waste Heat Recovery System for a Cement Manufacturing Process.*

Sultanov, M. M., Boldyrev, I. A., & Evseev, K. V. (2022). Application of machine learning to predict the thermal power plant process condition. *Journal of Physics: Conference Series, 2150*(1). https://doi.org/10.1088/1742-6596/2150/1/012029

Tyagi, K., Rane, C., Harshvardhan, & Manry, M. (2022). Regression analysis. In *Artificial Intelligence and Machine Learning for EDGE Computing* (pp. 53–63). Elsevier. https://doi.org/10.1016/B978-0-12-824054-0.00007-1

Investigation of Mechanical Properties and Terminological Performance of TiO_2 Filled Hybrid GFRP Composite and Nanocomposite—A Review

S. Prashanth, Sandeep Kumar Singh, Rosang Pongen, Yogendra Kumar Verma, and Kanjarla Vijay

Abstract Glass Fiber Reinforced Polymer (GFRP) composites have gained widespread adoption across various industries due to their outstanding strength-to-weight ratio, resistance to corrosion, and cost-effectiveness. However, their long-term reliability is often compromised by issues such as delamination and moisture absorption. This study investigates the improvement of mechanical properties and water resistance in GFRP composites through the integration of TiO_2 nanofillers. The epoxy matrix was reinforced with TiO_2 nanoparticles in concentrations ranging from 0.1 to 7 wt.%, with dispersion optimized using an Ultrasonication Dual Mixing (UDM) technique. Mechanical evaluations, including tensile, flexural, and impact tests conducted in accordance with ASTM standards, revealed notable enhancements at an optimal TiO_2 content of 1 wt.%. Higher concentrations resulted in nanoparticle agglomeration, leading to reduced performance. Water absorption experiments indicated that composites with 0.1 wt.% TiO_2 exhibited the lowest moisture uptake due to improved interfacial adhesion. The results confirm that TiO_2 nanofillers contribute to enhanced stiffness, fracture toughness, and overall durability of GFRP composites. These advancements make TiO_2 reinforced GFRP materials highly suitable for demanding applications in aerospace, automotive, and marine engineering. This research underscores the potential of nanofiller-modified GFRP composites as a high-performance material for structural applications requiring superior mechanical integrity and environmental resistance.

Keywords Glass fiber · GFRP composites · TiO_2 Nanofiller · Mechanical property · Water absorption

---

S. Prashanth · S. K. Singh (✉) · R. Pongen (✉) · Y. K. Verma · K. Vijay
Department of Mechanical Engineering, National Institute of Technology, Nagaland, India
e-mail: sandeepsingh15july@gmail.com

R. Pongen
e-mail: rosangpongen@nitnagaland.ac.in

M. Ram et al. (eds.), *Technological Advancements in Engineering and Management Sciences*, Management and Industrial Engineering,
https://doi.org/10.1007/978-3-032-04606-2_7

Nomenclature

TiO_2 (Titanium Dioxide)
GFRP (Glass Fiber Reinforced Composite)

1 Introduction

A composite material is a structural substance comprising two or more distinct components that remain immiscible while being combined at a macroscopic level. It primarily consists of two phases: the reinforcement phase—incorporating fibers, particulates, or flakes, and the matrix phase, which is a continuous medium typically composed of polymers, metals, or ceramics (Kaw 2006). The advancement of composite materials, integrating both natural and synthetic fibers such as glass, carbon, and aramid, has gained significant research attention. Owing to their superior mechanical characteristics, composites have become a focal point for scientists and engineers seeking innovative alternatives to conventional materials. Among the various reinforcement fibers, glass fiber has emerged as a widely preferred option for polymer composites due to its unique blend of lightweight characteristics, high strength-to-weight ratio, low density, cost-effectiveness, and ease of fabrication (Singh et al. 2024a). Fiber-reinforced polymer (FRP) composites offer notable advantages over traditional materials, including excellent impact resistance, enhanced damping properties, and improved toughness. These composites are extensively utilized across diverse industries, including automotive, aerospace, marine, sports equipment, transportation, and infrastructure applications (Erden et al. 2010a). The mechanical performance of fiber-reinforced composites is dictated by factors such as fiber strength, stiffness, chemical resistance, and alignment. Similarly, the matrix properties play a crucial role, depending on its inherent strength and the interfacial bonding with the reinforcement phase. Since the matrix constitutes the bulk of the composite, it significantly influences its shape and dimensional stability. It also functions to anchor the fibers, distribute loads efficiently, shield them from environmental degradation, and enhance resistance to impact, fracture, and transverse stress (Awan et al. 2009).The optimization of fiber orientation and composition is a key area in the development of glass fiber-reinforced polymer (GFRP) composites. These materials exhibit rigidity comparable to steel while being substantially lighter, offering nearly four times lower density than steel and superior structural integrity compared to aluminum (Alam et al. 2010). The exceptional mechanical attributes of GFRP composites result from the precise selection and strategic arrangement of reinforcing fibers within the polymer matrix. Various forms of glass fibers—such as continuous strands, short fibers, chopped fibers, and woven fabrics—are employed in composite fabrication, with chopped fiber mats particularly enhancing mechanical strength and wear resistance (Kaushal Kumar et al. 2016). The performance of these materials is primarily influenced by fiber orientation and matrix selection.

In composites reinforced with continuous fibers, a layered architecture is established, yielding superior mechanical behavior along the longitudinal axis, though the transverse properties remain relatively weaker (Goyat et al. 2011). One of the primary challenges in GFRP composites without nanofillers is delamination due to insufficient interfacial adhesion between the fiber and matrix, as observed by Kumar et al. (Halder et al. 2013). This limitation restricts their application in high-performance sectors such as aerospace, automotive, and sporting equipment. To address this issue, researchers have explored the integration of nanofillers into epoxy matrices, leading to remarkable enhancements in mechanical properties. The incorporation of organic and inorganic nanofillers has effectively mitigated delamination concerns, enhancing the structural stability of GFRP laminates. Various inorganic nanoparticles, including aluminum oxide (Al_2O_3) (Sandeep Kumar Singh and Thingujam Jackson Singh 2022), SiO_2 (Halder et al. 2012), TiO_2 (Singh et al. 2024b), zirconium dioxide (ZrO_2) (Singh et al. 2023), MWCNTs, and Graphene and its derivative such as GO and rGO (Chen and Mao 2007), have been successfully introduced into polymer matrices, resulting in significant performance enhancements and graphene (Singh et al. 2023), have been explored for this purpose. Among these, titanium dioxide (TiO_2 has attracted significant interest due to its distinctive attributes, including a high refractive index, non-toxicity, resistance to corrosion, affordability, self-cleaning ability, antibacterial nature, dielectric properties, and catalytic performance (Chen and Mao 2007). Recent studies have demonstrated that the addition of nanofillers such as TiO_2 can substantially minimize delamination in fiber-reinforced polymer (FRP) composites. When these nanofillers undergo appropriate surface modification, they improve adhesion at the fiber-matrix interface, consequently reducing manufacturing and production expenses (Zhai et al. 2015). In the present investigation, another crucial factor is evaluated—specifically, the water absorption behavior of GFRP laminate composites under controlled environmental conditions. Composite materials are often exposed to a variety of harsh environments, including water, hydrothermal conditions, extreme temperatures (both low and high), alkaline and corrosive media, as well as ultraviolet (UV) radiation. In hydrothermal environments, polymer matrices tend to absorb moisture, which can be classified into two types: bound water and free water. Bound water is chemically attached to the hydroxyl groups in the epoxy matrix, whereas free water accumulates as clusters at the interface between the matrix and the fibers (Visco et al. 2008; Nayak et al. 2016). This absorbed moisture can compromise the composite's physical, chemical, electrical, and mechanical properties. Specifically, the epoxy may undergo physical changes such as plasticization and swelling, and chemical alterations like chain scission and hydrolysis (Singh et al. 2025; Sridhar I and Venkatesha CS. Variation of damping property of polymer composite under saline water treatment 2013). Moreover, under hydrothermal conditions, differential swelling and osmotic effects can lead to micro-crack formation within the matrix due to interfacial debonding (Balu Maloth et al. 2020). One promising approach to mitigate these issues is the incorporation of TiO_2 nanofillers to seal the inherent pores and voids within the composite structure. By employing ultrasonication, the TiO_2 nanofillers can be uniformly dispersed within the epoxy resin (Vishnu Prasad et al. 2018). This improved distribution efficiently occupies the micro-gaps, strengthening the attachment between the reinforced fiber and the polymer matrix. Consequently, the likelihood of swelling, osmotic fractures, and moisture infiltration is considerably minimized, enhancing the composite's overall longevity (Aramide et al. 2012).

This study investigates the improvement of GFRP composites by integrating TiO_2 nanofillers, emphasizing their influence on mechanical strength and moisture resistance in hydrothermal environments. The research underscores the importance of optimized Ultrasonication Dual Mixing (UDM) for uniform dispersion of TiO_2 nanoparticles, which enhances mechanical interlocking among the matrix and fibers. The impact of different TiO_2 nanofiller weight fractions on tensile, flexural, and impact properties is analyzed, revealing that 1 wt.% TiO_2 provides the best mechanical performance, whereas excessive loading causes particle agglomeration, diminishing effectiveness. Additionally, the presence of TiO_2 nanoparticles minimizes void formation, leading to lower water absorption and greater structural longevity. These improvements make nanofiller-reinforced GFRP composites highly suitable for aerospace, automotive, and marine applications, where a high strength-to-weight ratio and environmental resilience are essential.

2 Materials and Methods

2.1 Classification of Synthetic Glass Fiber

Fibers serve as a fundamental component in the fabrication of FRP composites. This research specifically focuses on synthetic glass fibers, as illustrated in Fig. 1. Additionally, the chemical composition of various types of glass fibers is presented in Table 1. The primary categories of glass fibers include.

A-glass: A-glass, also known as alkali glass, exhibits strong chemical resistance. Its composition closely resembles that of window glass, making it a durable and versatile material. In certain regions, A-glass fiber is utilized in the manufacturing of process equipment due to its reliable properties and structural integrity.

C-glass: Commonly referred to as *chemical glass*, C-glass is highly resistant to chemical degradation, making it suitable for applications requiring excellent chemical stability.

D-glass: This type of glass consists of borosilicate material with a low dielectric constant, making it ideal for electrical insulation purposes.

E-glass: Also known as *electrical glass*, E-glass is widely used as an excellent electrical insulator due to its superior insulating properties.

AR-glass: Designed to withstand alkaline environments, AR-glass exhibits high resistance to alkali-induced deterioration.

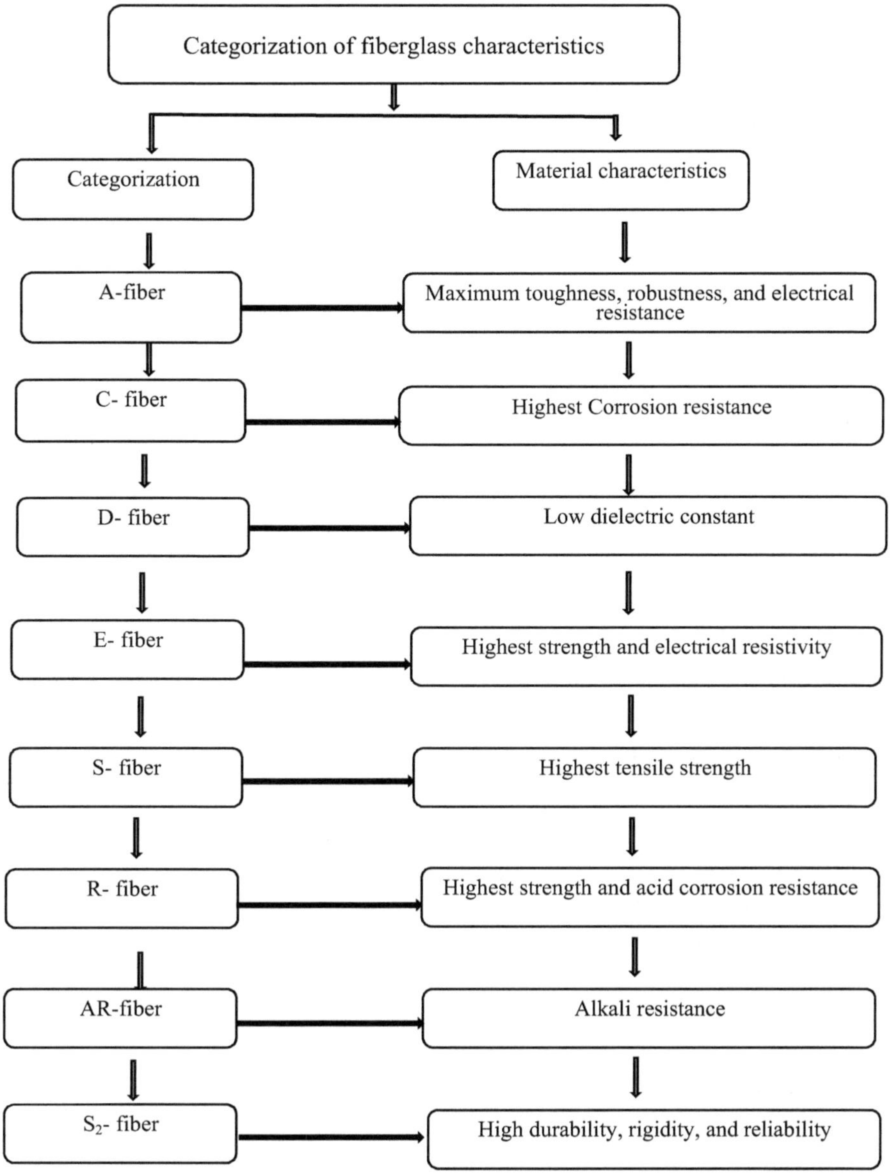

Fig. 1 Categorization of fiberglass characteristics (Sathishkumar et al. 2014)

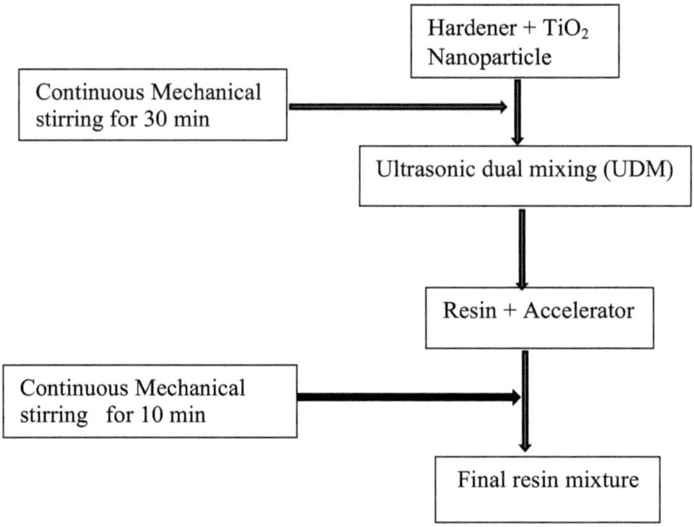

Fig. 2 Flow chart for preparation of resin mixture with nanofiller

R-glass: Composed of calcium aluminosilicate, R-glass is primarily utilized for reinforcement in applications requiring enhanced strength and resistance to acid corrosion.

S-glass: Commonly referred to as *structural glass*, S-glass is well known for its outstanding mechanical properties, making it a preferred choice for high-performance applications.

S2-glass: This type of glass is formulated with a magnesium aluminosilicate base and is primarily utilized for reinforcing textiles or in composite structural applications that demand exceptional strength, rigidity, and resistance to high temperatures and harsh environments.

2.2 *Fillers*

Nanofiller mainly indicates those filler with at list one of its dimensions in nanolevel. Filler should be non-reactive, non-corrosive, dry with uncured resin. It can enhance cross linking density of epoxy resin. Secondary organic or inorganic nanofiller or dispersed either on surface of fiber or in epoxy matrix to reduce delamination of GFRP composite. The integration of nanofillers into polymer-based composites enhances their mechanical performance while simultaneously lowering production expenses. These nanomaterials contribute to greater matrix rigidity, enhanced resistance to crack propagation, and improved fracture toughness. However, the degree of these enhancements is influenced by multiple factors, such as the type

Table 1 Preparation of nanocomposite and GFRP composite by various technique

S. no	Fabrication technique	Process	Ref.
1	Hand lay-up method process by compression moulding	The woven roving mat E-glass fiber, with or without filler, was used to fabricate a polyester matrix composite through the hand layup method, followed by compression molding at a pressure of 120 bar at ambient temperature for a duration of 120 min. The composite laminate had a glass fiber content of 37% by volume and a thickness of 3.5 mm	Erden et al. (2010b)
2	Hot press technique	The laminate was fabricated by shearing orthogonal fabric into a non-orthogonal form at various weaving angles. Epoxy and hardener (10:3 wt. ratio) were mixed, and the composite underwent pre-curing at 60 °C for 2 h, followed by curing at 93 °C for 4 h under 0.35 MPa pressure	Atas and Liu (2008)
3	Compression moulding	The fiber sheets were cut to the required dimensions and dried in a hot air oven at 150 °C to eliminate any moisture. The epoxy resin and curing agent were blended in the correct weight ratio, and the pre-measured fiber layers were infused with the epoxy to achieve a final composite thickness of 3 mm. The laminate was then positioned in a mold and subjected to an initial curing process at 180 °C for 3 h, followed by a secondary curing stage at 200 °C for 2 h. Lastly, the composite was allowed to cool gradually to ambient temperature	Aktas et al. (2009)

(continued)

Table 1 (continued)

S. no	Fabrication technique	Process	Ref.
4	H- Type press	A woven glass fiber (GF) reinforced epoxy composite was fabricated using oil cake filler. GF (18 μm diameter) mats were impregnated with an epoxy-hardener mixture (100:38 wt.%) via magnetic stirring and roller-brush application for uniform wetting	Mohan et al. (2011)
5	Hand lay-up method process by hydraulic press	A resin-to-hardener weight proportion of 100:12 was maintained, and the blend was combined with the reinforcing fibers to achieve a final laminate thickness of 3 mm. To ensure proper consolidation, the laminate was subjected to a hydraulic press at a pressure of 0.5 MPa After the initial fabrication, the specimen underwent a pre-curing stage at ambient temperature for 24 h. Once this process was completed, the sample was carefully extracted from the mold and subsequently post-cured in an electric oven at 120 °C for 2 h to improve its mechanical and thermal characteristics. The final dimensions of the prepared composite laminate were 250 × 250 × 3 mm3	Suresha et al. (2006)

of nanofiller, its particle size, morphology, concentration, dispersion quality, and compatibility with other composite constituents. In this investigation, various inorganic nanofillers including aluminum oxide (Al_2O_3), SiO_2, TiO_2 zirconium dioxide (ZrO_2), MWCNTs, clay, and GO were employed, as documented in previous studies (http 2014; Goyat et al. 2017). Achieving homogeneous nanofiller dispersion is a crucial aspect of optimizing the performance of GFRP composites, as it promotes strong interfacial attachment between the fiber reinforcement and the matrix. Improved dispersion improves the filler-matrix interface, allowing for more effective stress transfer from the matrix to the fibers. In this study, TiO_2 nanofillers were introduced into an epoxy-based matrix, significantly improving both tensile and flexural characteristics, further validating their role in reinforcing composite materials (Bittmann et al. 2011).

2.3 *Matrix*

The matrix plays a crucial role in composite materials, as it forms the primary structure that defines their shape and size. Its main function is to attachment the fibers together while facilitating load transfer between them. Additionally, the matrix serves several key purposes, including shielding the fibers from environmental factors, enhancing the composite's resistance to impact and fracture, and improving the transverse properties of laminates. Depending on the application, matrices can be classified into different categories, such as organic matrices, metal matrices, and ceramic matrices.

2.4 *Thermosetting Polymer*

Thermosetting polymer matrix materials offer several advantages, including their ability to be processed with low-melting-point materials and their ease of mixing with reinforcing fibers before curing. These polymers undergo a chemical reaction that creates permanent crosslinked bonds, making them highly durable. However, one major drawback is their irreversible nature, meaning they cannot be remelted or reshaped after curing.

Despite this limitation, thermosetting polymers exhibit excellent mechanical strength, enhanced chemical resistance, and superior thermal stability, all while maintaining cost-effectiveness. Their structural integrity makes them ideal for various applications requiring durability and heat resistance.

Commonly, thermosetting polymers are used in composite manufacturing through methods such as hand lay-up and spray-up processes. Some widely used examples include epoxy, phenolic, polyester, and polyimide resins.

2.5 Thermoplastic Polymer

In recent years, thermoplastic matrix materials have gained more popularity over thermosetting polymer matrices due to their faster processing and curing times. Thermoplastic polymers soften upon heating and become fluid when exposed to higher temperatures. Unlike thermosetting polymers, their curing process is reversible, as no permanent chemical bonds are formed. Some common examples of thermoplastic polymers include polypropylene, polyamide, polyethylene, nylon, polycarbonate, and polystyrene.

2.6 Epoxy Resin

Epoxy is a thermosetting polymer, is one of most commonly used resin in composite industries. Epoxies do not give out byproduct during curing and hence have very low shrinkage during curing. Epoxies are well known containing epoxide group for there is good resistance to heat and chemical, excellence adhesive properties and good insulating properties. Epoxy resin are regularly used in application like coating, adhesive, encapsulate, casting materials, binding, plating materials, etc.

3 Methods

3.1 Preparation of GFRP Composite

Glass fiber reinforced polymer composite was prepared by adopting various technique such as:

- The manual lay-up technique followed by compression molding.
- Hot press technique
- Compression moulding
- H- Type press
- Hand lay-up process followed by hydraulic press

3.2 Preparation of Resin Solution with Nanofiller

Create a flowchart illustrating the process of mixing epoxy resin for fabricating a GFRP composite with TiO_2 nanoparticles as a filler. The formulation includes resin, hardener, and an accelerator. The resin-to-hardener weight ratio is 1:0.95, while the accelerator is added at 2weight percentage of the resins. First calculate amount of hardener and nanofiller put into a single hard plastic glass, here taken nanofiller 0.2,

0.4, 0.5 g respectively (Chang and Chow 2010). The whole mixture was put under overhead stirrer, the stirring action at 2000 rpm was continued up to 20 min. The slurry consisting of epoxy resin and TiO_2 nanoparticles was subjected to an ultrasonication dual mixing (UDM) process to ensure effective nanoparticle dispersion. This process involved simultaneous ultrasonication and mechanical stirring using an impeller for 15 min, following an intermittent pulse cycle of 10 s on and 30 s off. This approach facilitated the uniform integration of nanoparticles into the resin matrix. Subsequently, an accelerator was added to the resin mixture, followed by mechanical stirring at a speed of 3000 rpm for 10 min. As a result, a highly dispersed epoxy resin mixture was obtained (Goyat et al. xxxx). For optimal dispersion of nanoparticles in the viscous epoxy resin, the applied sonication energy must be sufficient to break nanoparticle agglomerates while preventing any degradation of the epoxy's molecular structure. During ultrasonication, localized heating occurs at the probe immersed in the epoxy resin, which can potentially alter the base material's properties. To mitigate this issue, cooling techniques such as placing ice cubes around the solution or using a chiller to circulate water were employed to regulate the temperature throughout the process. In addition, proper dispersion can be takes place (Chandramohan and Marimuthu 2010; Mishra et al. 2003; Chatterjee and Islama 2008).

3.3 *Mechanical Properties*

In this study, two key mechanical properties flexural strength and tensile strength along with their corresponding moduli, were evaluated to assess the stiffness and strength of the materials. Tensile strength was tested following ASTM D-3039 standards, while flexural strength was determined according to ASTM D-790. Additionally, impact strength was measured based on ASTM D-256 standards. Prior to testing, rectangular samples were precisely cut from composite sheets using a sharp razor blade, and the outer surfaces were polished using 1200-mesh emery paper to ensure a smooth finish (Prashanth et al. 2024). It is well established that the functional characteristics GFRP composites are comparable to steel in terms of strength, and their stiffness exceeds that of aluminum, while their specific gravity is only about a quarter of that of steel (Chisholm et al. 2004). However, when using neat epoxy to fabricate GFRP composites, delamination between layers can occur due to poor interfacial bonding between the matrix and the fibers (Halder et al. 2013). The introduction of TiO_2 nanofillers into the GFRP composite significantly enhances its tensile, flexural, and impact strength (Qiao et al. 2019). In the present research, a comparison was made between nanocomposites and GFRP composites both with and without nanofillers, focusing on the physical and mechanical properties of numerous glass fibers and TiO_2 nanofillers.

3.4 Tensile Properties

From Table 2, it is evident that all glass fibers possess specific properties; however, the inclusion of TiO_2 particles enhances these properties more effectively compared to other glass fibers. The incorporation of nanofillers into epoxy composites significantly improves mechanical characteristics, such as tensile and flexural strength. As shown in Table 3, and illustrated in Fig. 3. The experimental data indicate that epoxy modified with 1 wt.% TiO_2 nanoparticles achieves a yield strength of 10.3417 MPa and a maximum tensile strength of 35.3505 MPa. This improvement is attributed to the high dispersion of TiO_2 nanoparticles within the epoxy resin and amine hardener matrix, ensuring effective stress transfer and reinforcing the composite. However, at higher TiO_2 loadings (3, 5, and 7 wt.%), the mechanical properties decline compared to the 1 wt.% TiO_2 composite, although they still surpass those of pure epoxy. This decrease in strength at maximum concentrations is primarily due to increased particle-to-particle interactions, which lead to nanoparticle agglomeration. This agglomeration disrupts the uniform dispersion of particles, reducing the composite's overall mechanical performance. (Annlin Bezy et al. 2015). On the other hand, at higher wt.% of TiO_2 nanoparticle is lower value of tensile strength due to enlargement in matrix system. Micro filler commonly increases the stiffness of matrix but on other side they may some detrimental effect on the flexural strain to break. But in case of nanofiller no such type of issue because it has high specific surface area i.e., more reactive to other material and lower agglomeration (Advances in Nanocomposites - Synthesis, Characterization and Industrial Applications‖- Polymer Nanocomposite Materials for Structural Applications, Vijaya Rangari 2008).

The interface quality in composite materials, especially the static adhesion strength and interfacial stiffness, is critical for efficient load transfer and enabling elastic deformation between the matrix and the filler. In epoxy-based nanocomposites, the tensile properties initially improve with increasing ultrasonic dispersion machining (UDM) amplitude, with a maximum enhancement of up to 60%. However, at higher UDM amplitudes, there is a notable decline in these properties, decreasing by as much as 70%. The optimal tensile strength occurs at a 60% UDM power amplitude, showing an approximate 22% improvement over neat epoxy adhesives. Conversely, at 70% amplitude, the tensile properties diminish due to excessive heat generated near the ultrasonic horn. This heat causes the epoxy matrix to degrade, which negatively affects the toughness and strength of the composite material (Kaushal Kumar, et al. 2016).

3.5 Flexural Properties

The flexural test measures the force necessary to bend a material using a three-point loading system. Meanwhile, the flexural modulus represents the material's stiffness or resistance to bending. It is calculated as the ratio of stress to strain during the

Table 2 Properties of glass fibers and TiO_2 nanofiller

Fibers	Density (g/cm3)	Tensile strength (GPa)	Young's modulus (GPa)	Elongation (%)	Coefficient of thermal expansion (10^{-7} / °C	Poison's ratio	Refractive index	Reference
A-glass	2.44	3.310	68.9	4.8	73	–	1.538	http: (2014)
C-glass	2.52	3.310	68.9	4.8	63	–	1.533	
D-glass	2.11–2.14	2.415	51.7	4.6	25	–	1.46	
E-glass	2.58	3.445	72.3	4.8	54	0.2	1.558	
S_2-glass	2.46	4.890	86.9	5.7	16	0.22	1.558	
R-glass	2.54	4.135	85.5	4.8	33	–	1.546	
AR-glass	2.70	3.241	73.1	4.4	65	–	1.562	
EGR-glass	2.72	3.445	80.3	4.8	59	–	1.579	
TiO_2	4	51.6	288	–	-	0.27	2.7	Resource et al. (xxxx)
Epoxy	1.15	0.070	3.6	1.8	1.0–1.3	0.030	1.50–1.56	Rathore et al. (2016)

Table 3 Compare the mechanical properties of experimental data with filler without filler (Annlin Bezy et al. 2015)

Material	Yield strength (MPa)	Tensile strength (MPa)	Maximum modules (GPa)	Flexural strength (MPa)	Flexural modules (GPa)	% Elongation
Pure Epoxy (E)	4.10	8.48838	1.9335	95.13	2.6262	1.4582 (1.5–3.6)
E + 1wt% TiO_2	10.3417	35.35057	1.6584	80.67	2.8726	2.8498
E + 3wt% TiO_2	7.6536	21.32218	1.8048	57.84	2.8877	1.3916
E + 5wt% TiO_2	5.9775	28.82103	1.9069	53.06	3.3819	1.8388
E + 7wt% TiO_2	6.5103	24.3175	1.7630	78.21	3.2190	1.7832

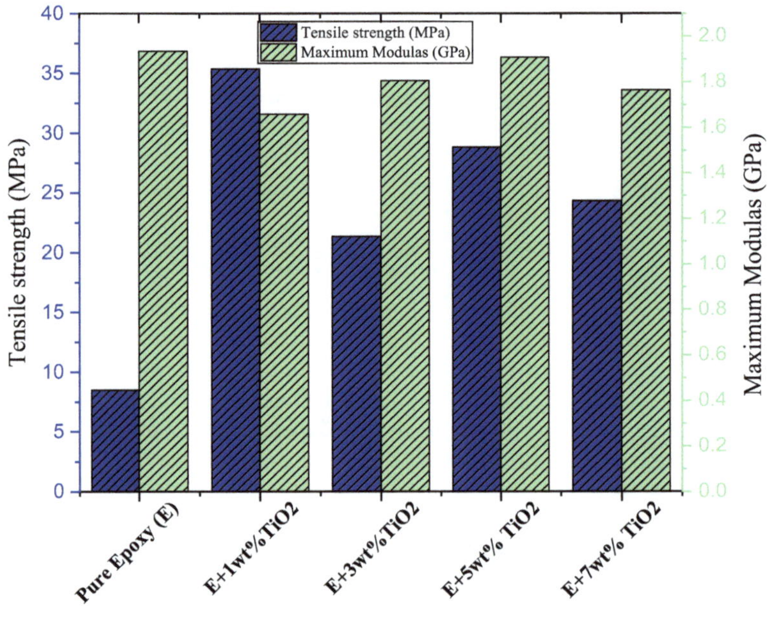

Fig. 3 Tensile properties of different weight percentages of TiO_2 Filler

initial deformation phase, in accordance with the ASTM D-790 standard. Glass fiber-reinforced polymer (GFRP) composites without nanofillers exhibit moderate flexural properties; however, their structural applications are restricted due to delamination. This issue primarily arises from inadequate mechanical inter locking among the epoxy resin and the reinforcing fibers. To overcome these limitations, titanium dioxide (TiO_2 nanofillers are incorporated into the matrix. The uniform dispersion

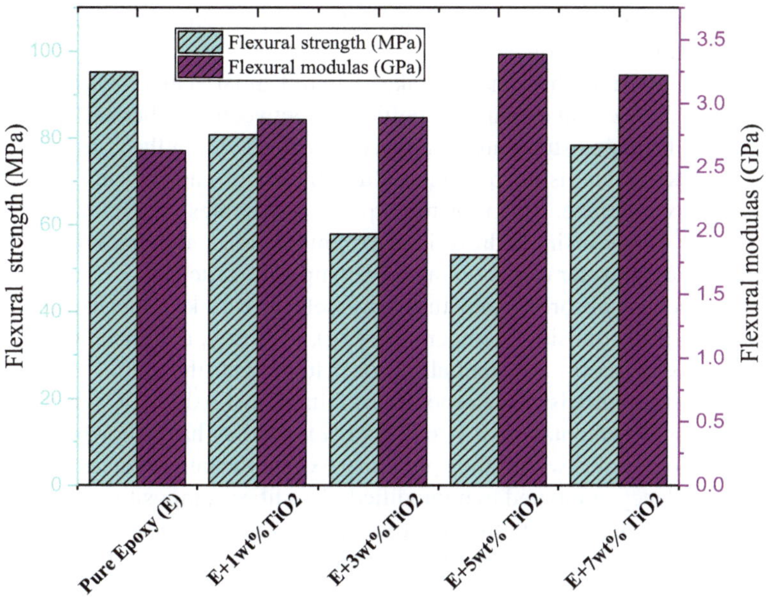

Fig. 4 Flexural properties of different weight percentages of TiO_2 Filler

of TiO_2 nanoparticles, achieved through ultrasonic dispersion (UDM), significantly improves resin cross-linking, reducing brittleness and enhancing the better interfacial bonding between the TiO_2 filler and matrix. Consequently, delamination is minimized, leading to an overall improvement in mechanical properties. As demonstrated in Table 2 and illustrated in Fig. 4, pure epoxy exhibits a lower flexural modulus of 2.6262 GPa, while epoxy reinforced with 5 wt.% TiO_2 nanofiller achieves a higher modulus of 3.3819 GPa. The highest flexural strength is observed at 7 wt.% TiO_2 loading, which can be attributed to the plasticizing effect of the filler within the epoxy matrix (Singh and Halder 2024). However, beyond this concentration, an uplift the filler concentration leads to a decrease in flexural strength. This reduction occurs due to the increased constraint on polymer chain mobility, causing a shortening of chain segments below the critical length. As a result, the ability of the matrix to sustain stress diminishes. Nonetheless, the flexural strength of the nano-filled epoxy remains superior to that of neat epoxy, which relies on weak van der Waals interactions, albeit in large numbers, to maintain structural integrity. (Al-Ajaj et al. 2013).

3.6 *Water Absorption*

Moisture absorption has a significant effect on the mechanical properties of glass fiber-reinforced polymer (GFRP) composites. Water infiltrates the composite material primarily through pores and voids, which are often introduced during mechanical stirring. The moisture diffusion occurs through the matrix and along the fiber-matrix interface via capillary action.

As reported by Ramesh Kumar Nayak, Kishore Kumar et al. (Nayak et al. 2016), a mismatch in the thermal expansion coefficients between the fiber and matrix interphase leads to localized stress and strain accumulation within the composite. Additionally, the absorbed moisture propagates further through capillary movement along the fiber-matrix interface. This moisture uptake contributes to swelling and plasticization of the epoxy resin, hydrolysis of polymer chains, and crack formation, all of which accelerate water diffusion into the composite structure.

The presence of absorbed moisture adversely affects key mechanical properties, including glass transition temperature (Tg), viscosity, and the elasticity of the polymer (Won et al. xxxx). A potential approach to mitigate these effects and enhance mechanical performance is the incorporation of titanium dioxide (TiO_2) nanoparticles at varying weight percentages (0.1 to 0.7 wt.%). Studies have shown that embedding TiO_2 nanoparticles within the epoxy matrix significantly reduces the moisture absorption tendency compared to unmodified glass fiber composites. This reduction in water uptake is attributed to the effective filling of micro voids by the nanofillers. However, an excessive increase in TiO_2 content can lead to a higher degree of particle agglomeration, potentially reintroducing voids in the composite. The optimal reduction in water absorption was observed at 0.1 wt.% TiO_2, likely due to the formation of a well-integrated interface between the matrix and filler. The diffusion coefficient showed an approximate 9% reduction in TiO_2 modified composites compared to untreated glass fiber composites (Chatterjee and Islam xxxx).

3.7 Application

With the rise of global environmental awareness, there is a growing emphasis on utilizing composite materials like glass fiber, carbon fiber, and aramid fiber as reinforcements in polymer-based matrices. Nevertheless, the present research is specifically centered on GFRP composites and their nanocomposite variants. It is important to note that composites fabricated using the same reinforcement system may not always exhibit optimal performance under varying loading conditions. This is primarily due to inadequate interfacial bonding between the matrix and the reinforcing fibers, which affects the overall mechanical properties and durability.

According to Sanjay M. Ra, Arpitha G., and B. Yogesh et al. (Ra et al. 2015), the incorporation of titanium dioxide (TiO_2) nanofillers has been shown to enhance both the mechanical performance and moisture resistance of composite materials. Various types of glass fibers and polymer matrices are employed in composite manufacturing, and their specific compositions determine their suitability for different applications, with or without nanofillers. Glass fiber-based composites are widely utilized across diverse industries, including consumer goods, low-cost housing, aerospace, sports equipment, and civil infrastructure. The relatively high cost of conventional lightweight reinforced plastics has limited their widespread adoption, making GFRP an attractive alternative (Sathishkumar et al. 2014). Several advantages include low maintenance, a high strength-to-weight ratio, non-conductivity, superior corrosion resistance, excellent impact resistance, and cost-effectiveness. Additionally, these materials help mitigate electrical hazards, exhibit fire-retardant properties, and allow easier installation due to their lightweight nature. The applications of GFRP composites and their nanocomposite counterparts extend across multiple sectors, including aerospace, sports, electronics, and the automotive industry, where their unique properties deliver substantial performance benefits.

Electronics GFRP has been widely used in circuit board manufacturing (PBC), TV, frame of refrigerator, radio, electric motor cover, cell phone, computer, etc.

Aviation and aerospace Glass fiber reinforced polymer (GFRP) has found extensive use in the aviation and aerospace industries, although it is not commonly employed for primary airframe structures like jet engines, propellers, and exhaust fans, due to the availability of alternative materials that are more suitable for these specific applications. However, GFRP is widely utilized in various secondary components, including engine cowlings, luggage compartments, etc.

4 Conclusion

This research highlights the enhancement of mechanical properties and water absorption resistance in GFRP composites and nanocomposites through the incorporation of TiO_2 nanofillers. The findings demonstrate that:

- The use of pure epoxy resin matrix leads to reduced mechanical performance due to delamination issues. However, these challenges are effectively mitigated by the addition of TiO_2 nanofillers, which enhance adhesion and structural integrity.
- Surface modification of TiO_2 nanoparticles, combined with ultrasonic dual mixing (UDM), significantly improves dispersion, homogeneity, and better interfacial bonding, leading to superior mechanical properties.
- The addition of TiO_2 nanofillers significantly improves essential mechanical properties such as tensile strength, modulus, fracture toughness, fatigue resistance, and flexural strength, even at minimal concentrations.
- Optimal improvement is observed at 1 wt.% TiO_2 content, while excessive loading leads to agglomeration, reducing polymer interaction and mechanical performance.
- A minimal addition of 0.1 wt.% TiO_2 reduces the moisture diffusion coefficient by approximately 9%, demonstrating improved durability in moisture-prone environments.

These results emphasize the suitability of TiO_2 reinforced GFRP composites for advanced applications in sectors like aerospace, automotive, and marine engineering. Their improved mechanical durability and resistance to environmental factors make them ideal for demanding operational conditions.

References

Advances in Nanocomposites - Synthesis, Characterization and Industrial Applications‖- Polymer Nanocomposite Materials for Structural Applications, Vijaya Rangari, https://doi.org/10.5772/15615, 2008.

Aktas M, Atas C, Icten BM, et al. An experimental investigation of the impact response of composite laminates. Compos Struct 2009; 87: 307–313.

Ikram A. Al-Ajaj, Muhannad M. Abd and Harith I. Jaffer, —Mechanical Properties of Micro and Nano TiO2/Epoxy Composites‖, International Journal of Mining, Metallurgy & Mechanical Engineering (IJMMME) Volume 1, Issue 2; ISSN 2320–4060, 2013.

Alam S, Habib F, Irfan M, et al. Effect of orientation of glass fiber on mechanical properties of GRP composites. J Chem Soc Pak 2010; 32: 265–269.

N.Annlin Bezy, A.Lesly Fathima et.al. International Journal of Engineering Research and General Science Volume 3, Issue 5, September-October, 2015 ISSN 2091–2730 143 .

Aramide FO, Atanda PO and Olorunniwo et al. 2012. Mechanical properties of a polyester fiber glass composite. Int J Compos Mater 2012; 2: 147–151.

Atas C and Liu D. Impact response of woven composites with small weaving angles. Int J Impact Eng 2008; 35: 80–97.

Awan GH, Ali L, Ghauri KM, et al. Effect of various forms of glass fiber reinforcements on tensile properties of polyester matrix composite. J Faculty Eng techno 2009; 16: 33–39.

Balu Maloth, N.V. Srinivasulu et al. 2020 nfluence of titanium oxide fillers on the tensile and flexural properties of E-glass fabric/epoxy composites, https://doi.org/10.1016/j.matpr.2020.01.462.

B. Bittmann, F. Haupert and A. K. Schlarb, Ultrason. - Sonochemistry, 2011, 18, 120– 12

Chandramohan D, Marimuthu K. Thrust force and torque in drilling the natural fiber reinforced polymer composite materials and evaluation of delamination factor for bone graft substitutes – a work of fiction approach. International Journal of Engineering Science and Technology: 2(10), 2010,

Chang LN, Chow WS. Accelerated Weathering on Glass Fiber/Epoxy/Organo-montmorillonite Nanocomposites. Journal of Composite Materials 2010; 44:1421–34. https://doi.org/10.1177/0021998309360944.

Amit Chatterjee, Muhammad S. Islama et al. Fabrication and characterization of TiO2–epoxy nanocomposite. Materials Science and Engineering A 487 (2008) 574–585.

Chatterjee A, Islam MS. Fabrication and characterization of TiO2–epoxy nanocomposite. Mater Sci Eng A 2008;487: 574–85. http://dx.doi.org/ https://doi.org/10.1016/j.msea.2007.11.052

Chen X, Mao SS. Titanium Dioxide Nanomaterials: Synthesis, Properties, Modifications, and Applications. Chem Rev 2007; 107(7): 2891-2959.

N. Chisholm, H. Mahfuz, V.K. Rangari, A. Ashfaq, S. Jeelani, Comp. Struct. (2004) (web published).

Erden S, Sever K, Seki Y, et al. Enhancement of the mechanical properties of glass/polyester composites via matrix modification glass/polyester composite siloxane matrix modification. Fibers Polymer 2010a; 11: 732–737.

Erden S, Sever K, Seki Y, et al. Enhancement of the mechanical properties of glass/polyester composites via matrix modification glass/polyester composite siloxane matrix modification. Fibers Polym 2010b; 11: 732–737.

Goyat MS, Ray S, Ghosh PK. Innovative application of ultrasonic mixing to produce homogeneously mixed nanoparticulate-epoxy composite of improved physical properties. Compos Part A: Appl Sci and Manuf 2011; 42(10): 1421-1431.

M. S. Goyat, S. Rana, S. Halder and P. K. Ghosh, Ultrason. Sonochem, 2017, 40, 861– 873.

M.S. Goyat, P.K. Ghosh et al. Impact of ultrasonic assisted triangular lattice like arranged dispersion of nanoparticles on physical and mechanical properties of epoxy TiO2 nanocomposite. https://doi.org/10.1016/j.ultsonch.2017.11.019.

Halder S, Ghosh PK, Goyat MS. Influence of ultrasonic dual mode mixing on morphology and mechanical properties of ZrO2 -epoxy nanocomposite. H Perfo Polym 2012; 24(4): 331-341

Halder S, Ghosh PK, Goyat MS, Ray S. Ultrasonic dual mode mixing and its effect on tensile properties of SiO2-epoxy nanocomposite. J of Adh Sci and Technol 2013; 27(2): 111- 124.

http://www.slideshare.net/abiramprince/glass-fibermanufacture-and-applications (accessed 2 February 2014).

Kaushal Kumar, P.K. Ghosh et al. Improving mechanical and thermal properties of TiO2-epoxy nano composite, Composites Part B (2016), https://doi.org/10.1016/j.compositesb.2016.04.080.

Autar K Kaw. Mechanics of Composite Materials. 2nd, ed. CRC PRESS Taylor & Francis Group 2006. pp.3.

Mishra S, Mohanty A K, Drzal L T, Misra M, Parija S, Nayak SK. Studies on mechanical performance of biofiber/glass reinforced polyester hybrid composites. Composite Science and Technology: 63, 2003, pp.1377–85.

Mohan N, Natarajan S, Kumaresh Babu SP, et al. Investigation on sliding wear behaviour and mechanical properties of jatropha oil cake-filled glass-epoxy composites. J Am Oil Chem Soc 2011; 88: 111–117

Ramesh Kumar Nayak, Kishore Kumar Mahato et.ai. Water absorption behavior, mechanical and thermal properties of nano TiO2 enhanced glass fiber reinforced polymer composites. https://doi.org/10.1016/j.compositesa. 2016.09.003

Prashanth, S., Pongen, R., & Verma, Y. K. (2024). Mechanical and tribological performance of NaOH-Treated bamboo fiber-reinforced recycled HDPE composites under fretting wear conditions. *Wood Material Science & Engineering*, 1–15.

J. Qiao, Y. Wu, L. Li, Effect of filler content on the stress relaxation behavior of fly ash/polyurea composites, Polymer Testing (2019), https://doi.org/10.1016/j.polymertesting.2019.106168, Article:106168

Sanjay M Ra, Arpitha G & B Yogesh et al. Study on Mechanical Properties of Natural - Glass Fibre Reinforced Polymer Hybrid Composites: A Review. Materials Today: Proceedings 2 (2015) 2959 – 2967.

Rathore DK, Prusty RK, Kumar DS, Ray BC. Mechanical performance of CNTfilled glass fiber/epoxy composite in in-situ elevated temperature environments emphasizing the role of CNT content. Compos Part Appl Sci Manuf 2016; 84:364–76. https://doi.org/10.1016/j.compositesa.2016.02.020.

Online Materials Information Resource - MatWeb n.d. [accessed June 27, 2016].

Sandeep Kumar Singh and Thingujam Jackson Singh. Study of Mechanical Properties and Water Absorption Behavior of TiO2 Nanofiller-Enhanced Glass Fiber-Reinforced Polymer Composites: A Review. Techno-societal 2022, https://doi.org/10.1007/978-3-031-34644-6_82, pp 799–809

Sridhar I and Venkatesha CS. Variation of damping property of polymer composite under saline water treatment. IJIET. 2013, 2: 42.

Singh, S. K., & Halder, S. (2024). Synthesis, characterization, and application of hydroxyapatite grafted over pristine charcoal/activated charcoal for removal of pb2+ ions from aqueous solution. *Materials Today: Proceedings*, *106*, 107-114.

Singh, S. K., Nayak, B., Singh, T. J., & Halder, S. (2023). Investigating the role of synthesized reduced graphene oxide and graphite micro-fillers on mechanical and fretting wear performance of glass fiber epoxy-based composite. *High Performance Polymers*, *35*(9), 946-962.

Singh, S. K., Singh, T. J., Nayak, B., Sonker, P. K., & Singh, M. A. (2024a). Analysis of the impact of exfoliated graphene oxide on the mechanical performance and in-plane fracture resistance of epoxy-based nanocomposite. *High Performance Polymers*, *36*(9-10), 487-507.

Singh, S. K., Singh, T. J., & Halder, S., Khan N. Islam, Investigation of mechanical and thermo-mechanical properties of dopamine-functionalized TiO2/Epoxy nanocomposites. Published polymer composite 2025, https://doi.org/10.1002/pc.29755

Singh, S. K., Jackson Singh, T., Singh, L. D., Sonker, P. K., & Mazumder, B. (2024b). Experimental study on the impact of hybrid GFRP composites with graphene oxide and silicon carbide fillers on mechanical and wear properties. *Journal of Adhesion Science and Technology*, 1–28.

Suresha B, Chandramohan G, Prakash JN, et al. The role of fillers on friction and slide wear characteristics in glassepoxy composite systems. J Minerals Mater Characterization Eng 2006; 5: 87–101.
Visco AM, Calabrese L and Cianciafara P. Modification of polyester resin base composites induced by seawater absorption. Compos Part A 2008; 39: 805–814.
TP Sathishkumar, S Satheeshkumar et al. Glass fiber-reinforced polymer composites – a review Journal of Reinforced Plastics and Composites 2014, Vol. 33(13) 1258–1275! The Author(s) 2014.
Vishnu Prasad, M.A. Joseph, et al. Investigation of mechanical, thermal and water absorption properties of flax fibre reinforced epoxy composite with nano TiO2 addition. Composites: Part A (2018), https://doi.org/10.1016/j.compositesa. 2018.09.031.
Won J-P, Yoon Y-N, Hong B-T, Choi T-J, Lee S-J. Durability characteristics of nano-GFRP composite reinforcing bars for concrete structures in moist and alkaline environments. Compos Struct 2012; 94:1236–42. http://dx.doi.org/ https://doi.org/10.1016/j.compstruct.2011.11.006.
Zhai W, Wu ZM, Wang X, Song P, He Y, Wang RM. Preparation of epoxy-acrylate copolymer@nano-TiO2 Pickeringemulsion and its antibacterial activity. Prog in Org Coat 2015; 87: 122-128.

A Comprehensive Analysis of the μ-EDM Process: Exploring the μ-EDM Variants, Parameters, Circuits and Dielectrics

Subhankur Dutta, Aditya Kumar Mishra, Kishore Debnath, and Amit Kumar Singh

Abstract The electrical discharge machining (EDM) process has many variants like die-sinking μ-EDM, wire μ-EDM, etc. These processes are pragmatic in micro fields to manufacture micro slots, micro holes, micro moulds, microelectromechanical systems (MEMS), etc. The μ-EDM process is significantly used for making smaller dimensional components. This method is superior with regard to power density, material removal efficiency and energy efficiency in comparison to conventional EDM process. Various input parameters of the μ-EDM process have been analyzed to study their influence on the machining characteristics for quality assessment while fabricating at the micro level. This chapter systemically presents and discusses the vital aspects of the μ-EDM process. The μ-EDM variants are discussed by highlighting the influence of the different μ-EDM parameters. The electrical and non-electrical parameters are discussed in-depth. The μ-EDM response characteristics are elaborated in a systematic way. The study of micro tools, workpieces, and dielectric fluids with their proficiencies is also discussed. The electrical circuit of the EDM method based on the resistance–capacitance (R–C) generator and transistor pulse generator with respective features are explicated. The research perspectives in μ-EDM are also briefly discussed.

Keywords μ-EDM · EDM variants · μ-EDM parameters · EDM circuits and dielectrics

S. Dutta · A. K. Mishra · A. K. Singh (✉)
Department of Mechanical Engineering, National Institute of Technology Nagaland, Dimapur Nagaland, India
e-mail: amit.kumar965@yahoo.com; amit.me@nitnagaland.ac.in

S. Dutta
e-mail: subhankur91@gmail.com

A. K. Mishra
e-mail: aditya.1989mishra@gmail.com

K. Debnath
Department of Mechanical Engineering, National Institute of Technology Meghalaya, Sohra, India
e-mail: kishoredebnath@nitm.ac.in

M. Ram et al. (eds.), *Technological Advancements in Engineering and Management Sciences*, Management and Industrial Engineering,
https://doi.org/10.1007/978-3-032-04606-2_8

1 Introduction

The manufacturing industry has reached a zenith with the advent of new developments in the field of fabricating micro-components. The sectors of application are widespread, including airspace, robotics, optics, biomedicine, and electronics, to name a few. The cornerstone for the development of the μ-EDM technique is rooted in the high demand for precise tolerance levels and excellent integrity of the surface, particularly in the machining of alloys of titanium and nickel, composite materials of tungsten-carbide, numerous superalloys, and ceramics. Miniaturization is a major requirement for various industries. The conventional machining processes lag behind because of the extreme hardness and toughness along with some superior metallurgical properties of these materials (Boral et al. 2019). The electrical discharge machining (EDM) of the enormously brittle and conducting substances is performed by repetitive generation of electrical discharges without mechanical intervention of force while the tool and workpiece are aided by a dielectric agent. The method is economical and successfully implemented in lower energy ranges through one of its variants called μ-EDM. The vibrations and chatter-related inconveniencies are absent while performing micromachining by the μ-EDM process. The manufacturing of intricate three-dimensional patterns, micro-cavity, and micro-moulds is done extensively with the help of the μ-EDM process (Jahan et al. 2011). There are a few notable differences between EDM and μ-EDM in terms of the tool dimensions, techniques for producing the micro tools, flushing methods, energy-supplying techniques for sparking and to control the gap between the tool and workpiece. A few salient comparative points are mentioned below:

(a) The plasma channel generated in μ-EDM has almost an analogous dimension to the micro-tool as compared to the traditional EDM method, wherein the tool size is quite larger than channel radii (Davim 2013).
(b) The μ-EDM process is unable to dissipate large thermal energies from the micro-gap owing to the use of the small tool (less than 1 mm) resulting in restricted utilization of supplied energy (Davim 2013; Raju & Hiremath 2016).
(c) The traditional EDM process has become less efficient due to tool wear and vibration. However, μ-EDM gives highly precise and accurate end products because the aforementioned factors are controlled (Davim 2013).
(d) The peak current is lower in μ-EDM (< 3 A) than in traditional EDM (> 3 A). Also, open-circuit voltage ranges from 10 to 120 V in μ-EDM as compared to traditional EDM (4 to 400 V) (Raju & Hiremath, 2016).
(e) The flushing mechanism in μ-EDM should be highly efficient because of the small micro-gap (< 10 μm). The electrode must be handled with caution to counter the effect of deflection on it created by low stiffness and small electrode-pressure-area (Katz & Tibbles, 2005; Raju & Hiremath, 2016).
(f) The wearing of the electrode in the μ-EDM is more than in traditional EDM (Davim, 2013).

(g) The crater size in μ-EDM is relatively diminished than traditional EDM because of restricting the extreme peak energy for controlling the MRR in every discharge (Uhlmann et al. 2005).

The chapter elucidates the operative philosophy of the μ-EDM technique by encompassing its variants with a short-term discussion of the pertinent literature. The input features and response characteristics are described in a methodical manner. The usage of tools, workpieces, and dielectrics in μ-EDM are elucidated. The μ-EDM circuits are demonstrated along with a comparison of the analytical aspects in micromachining. The μ-EDM scenarios for futuristic micromachining are briefly highlighted.

2 Working Philosophy of μ-EDM

The fundamental concept of material elimination in μ-EDM is like conventional EDM, with the tool as a cathode and the workpiece as an anode powered by a pulsed supply. The well maintenance of micro-gap and prevention of electrode short-circuit is monitored by a servo control. The dielectric setup comprises a reservoir, filter and pumping unit for supplying the freshened dielectric in the micro-gap and clearing out debris from it. The generation of an electrical field is initiated by the gap voltage resulting in breaking down the insulation behaviour of dielectrics. As the dielectric resistance drops down enormously, an individual discharge is formed and strikes the workpiece. Consequently, the vaporization and melting happen at the surface forming a crater. The intense thermal energy and workpiece-generated taints hinder the proper aligning of ion-converted dielectric molecules, increasing their resistivity. The dielectric cannot endure any stable sparking as the current falls and voltage increases. The pulse interval occurs as the current is turned off at this juncture. During the pulse-off time, the vapour cover engulfs the discharge bursts, creating a vacancy for the entry of freshened dielectric and debris elimination. The zone becomes cooler and ionization occurs again to develop favourable conditions for the ensuing discharge (Davim 2013). The schematic of μ-EDM methodology is depicted in Fig. 1.

3 Variants of μ-EDM

3.1 μ-EDM Drilling

In μ-EDM drilling, the machining of the workpiece is carried out with the help of a pipe-shaped electrode. To erode the workpiece, an electric current is used. In this process, the electrode doesn't tough the workpiece physically. As a result, the deflection or banding of the electrode is minimized as compared with the drilling

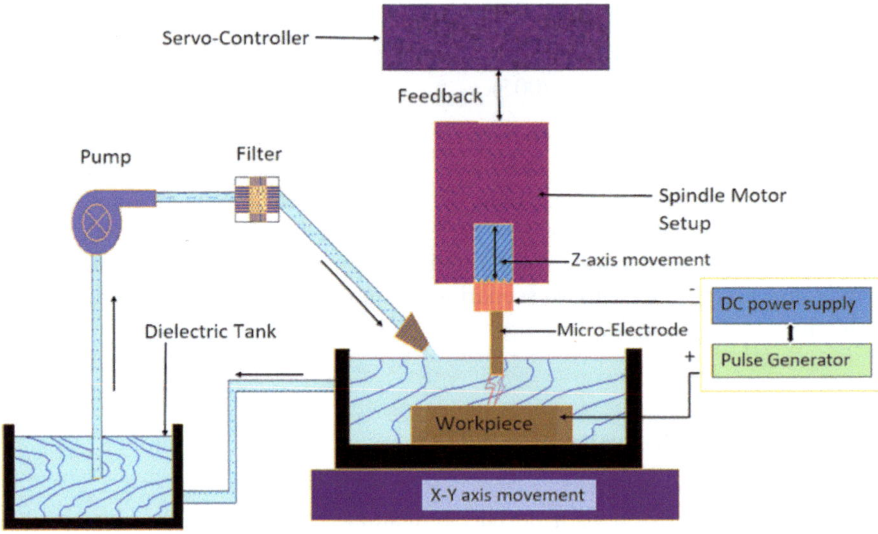

Fig. 1 Schematic representation of the μ-EDM process

holes by using conventional cutting tools. The electrode attains high speed during the drilling operation. A high-pressure flushing pump is used to flush out the scrap which is produced during the machining operation.

Figure 2 shows that the tool electrode movement is in a vertically downward direction. The tool feeds on the workpiece surface until the desired depth of the hole is achieved. The evaluation of MRR, TWR, overcut and taper angle (TA) was done on 316L-SS by finding the outcomes of polarity, capacitance, spindle speed and voltage in an R–C controlled μ-EDM drilling process. The analysis of variance (ANOVA) was conducted to find the effect and relationship of the input parameters on responses. The most influential parameters were voltage (9.23%), speed (14.30%), polarity (31.56%), and capacitance (34.85%). The optimum combination was a speed of 250 RPM, capacitance of 0.01 nF, and voltage of 100 V with a positive polarity for achieving low TA, overcut, TWR, and high MRR (Pilligrin et al. 2017). The drilling of metal glasses was done with great accuracy by using tool rotation as the key input parameter. The time for machining was greatly influenced by tool rotation reducing it by 5 to 10 times. The other response features viz. TA, edge deviation, and overcut were also affected by the tool rotation. The μ-EDM drilling exhibited more stability and low TA with an optimal combination of low capacitance, no tool rotation, and high voltage. (Kuriakose et al. 2020).

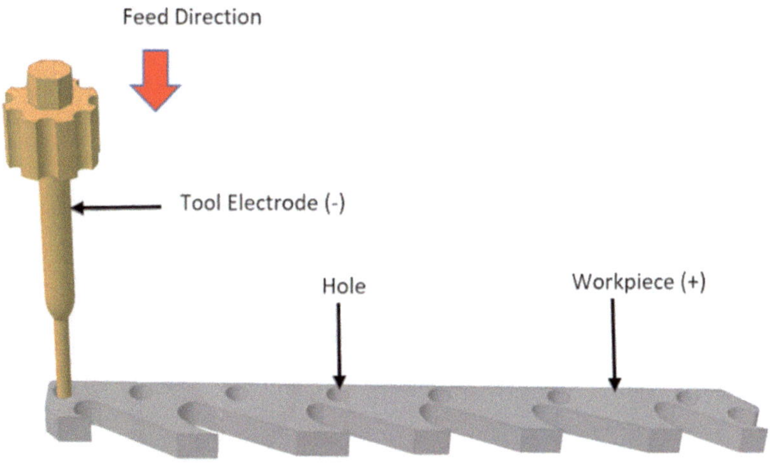

Fig. 2 Schematic representation of the μ-EDM drilling process

3.2 μ-EDM Milling

The μ-EDM milling is another variant of the μ-EDM process. Figure 3 represents the fundamental functioning of the μ-EDM milling method. It can be explained as a combined operation of the μ-EDM drilling along with the die-sinking technique. In μ-EDM milling, the tool is fed like a drill bit and travels in a horizontal x- and y-axis direction up to a predefined depth. The elimination of material is done by removing layers of the surface. The maximum thickness of the layer is up to 0.1 μm. However, the tool traverses on a simple tool path which is created in CAD/CAM software to machine complex three-dimensional microcavities. The machining of ceramic ($ZrC + 10MoSi_2$) and AISI-316L-SS was done by using μ-EDM milling for surface characterization by investigating skewness and surface roughness (SR) by studying the impact of discharge energy. A good surface finish was attained by combining high frequency and peak current with low pulse-on time and voltage for both materials. The distribution of skewness with machining times for both $ZrC + 10MoSi_2$ and AISI-316L-SS revealed the differences in their characteristics (Urso et al. 2018). The exploration of recast layer thickness (RLT) and other surface micro-features of Si_3N_4-TiN was done by μ-EDM milling. Also, TWR and MRR were assessed through the forms of various discharge pulses. The growth in normal pulses increased MRR and TWR. However, TWR decreased with delayed pulses and arcs. The final SR on the μ-EDM milling surface ranged from 0.77–0.98 μm because of the presence of micro-sized cracks and resolidified droplets (Marrocco et al. 2020).

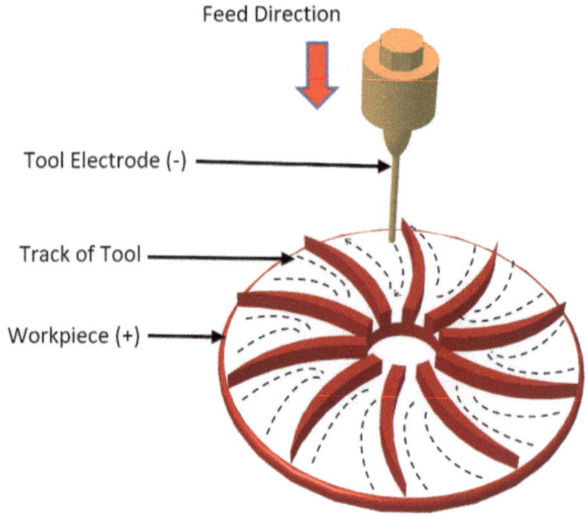

Fig. 3 Schematic representation of the μ-EDM milling process

3.3 *Die-Sinking μ-EDM*

The die-sinking μ-EDM is another type of μ-EDM which is very useful for the manufacturing of moulds and dies. The die-sinking μ-EDM is used to replicate the die (tool) shape on the workpiece by using electrical sparks, as shown in Fig. 4. An electrode (tool) is utilized to machine the workpiece in presence of a dielectric substance. The electrode wear is very low because the workpiece and tool do not touch each other. Compared to other types of μ-EDM, it is easier to handle, affordable, and less expensive. Also, the presence of static spindle, simple auxiliary systems, and the absence of complex axis controllers make the die-sinking μ-EDM more useful in the manufacturing sectors. The analysis of TWR, surface finish, and RLT for varying discharge currents was done for micro-cavity studies of EN-24 alloy steel using die-sinking μ-EDM. The gradual increase of current (1 to 2 A) raised the TWR (6.77 to 1.816 × 10^5 $\mu m^3/s$) and MRR (28.50 to 6.817 × 10^5 $\mu m^3/s$). The presence of cracks, voids, and surface deterioration was seen while micromachining performed at a current of 2 A, forming a thicker recast layer and rough surface (Vidya et al. 2021). The impact of discharge energy, capacitance, and voltage was assessed for evaluating TWR, overcut, and MRR of an EN-24 die steel during micro-hole fabrication. All the responses increased with a rise in gap voltage. The MRR increased with a capacitance range of 0.1 to 1 nF but dropped in the range of 1 nF to 10 nF. The TWR increased up to 1 nF capacitance but marginally decreased with a value of 10 nF and higher. It was also found that the discharge of energy aided the rise of TWR and MRR. The growth in discharge energy and capacitance triggered an increase in overcut (Kadirvel et al. 2014).

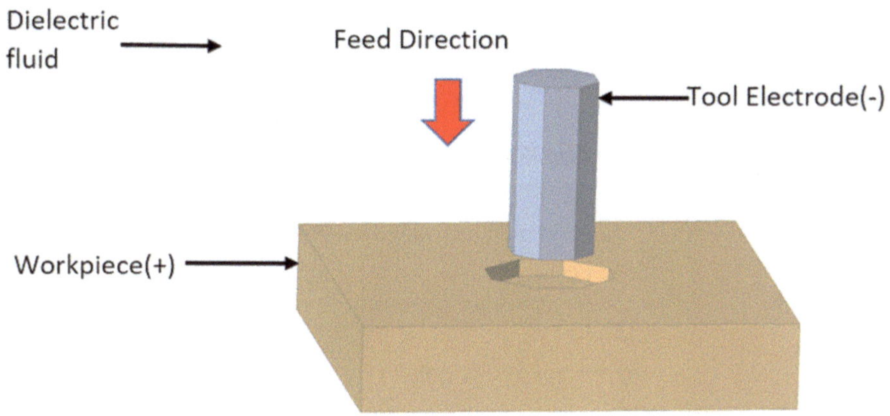

Fig. 4 Schematic diagram of the die-sinking μ-EDM process

3.4 *Reverse μ-EDM*

In reverse μ-EDM, the workpiece is connected to the cathode and the tool to the anode. Figure 5 shows that the electrode (tool) takes the shape of a thin sheet with multiple micro holes in the array form. The workpiece is a bulk circular rod connected to the spindle. The material elimination from workpiece occurs where the surface boundary of workpiece and tool surface are present. So, the replica of the hole is copied on the workpiece in the form of a thin cylindrical surface. In this way, the material is removed in the reverse μ-EDM method. This method is utilized in manufacturing micro-sized electrodes and cavities on the metal surface. The exploration of dimensional alteration, erosional rate, and machining duration was performed for several input combinations by using reverse μ-EDM during the machining of micro-rods of 182 μm in diameter. It was found that machining duration is reduced with a surge in voltage from 80 to 140 V and capacitance from 100 to 1000 pF. Also, the time for machining lessened with an upsurge in the feed rate from 5 to 20 μm/s. The rate of feed from 5–15 μm/s and voltage from 80–140 V surged the erosional rate. A high capacitance of 1000 pF lowered the rate of erosion. A combination of capacitance (1000 pF), voltage (120 V) and feed rate (5 μm/s) was utilized for fabricating micro-rod arrays from bulk with a diameter of 58 μm and length of 830 μm (Singh et al. 2019). The machining of a macro-rod-array structure using reverse μ-EDM in AISI-D2 steel was performed to find TWR, TA, and MRR. The pulse-on time highly affected the TWR, contributing 32.68%, followed by pulse current and gap voltage. The pulse current mostly impacted the MRR, contributing 49.07%, followed by gap voltage. The TA was influenced by pulse-on time which varied from 3 to 5° (Nikam et al. 2018).

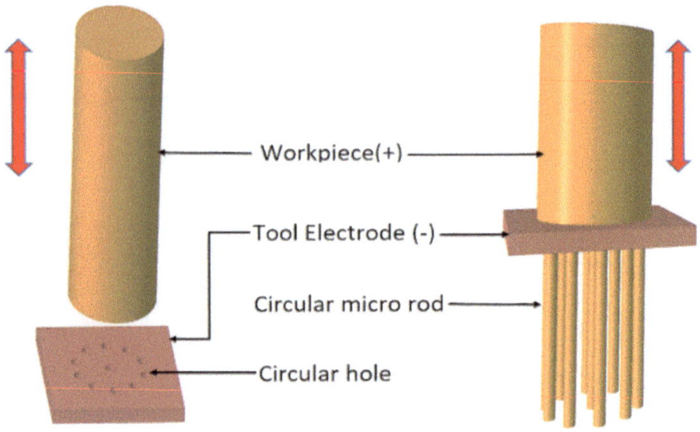

Fig. 5 Schematic representation of the reverse μ-EDM process

3.5 *Wire μ-EDM*

The wire μ-EDM is used to machine good metallic conductors. A continuous thin wire as a tool runs in a loop without coming into direct contact with a workpiece, as shown in Fig. 6. Two types of wire are used in a wire μ-EDM. The first is a consumable wire and the other one is a non-consumable wire. The consumable wire cannot be used again, but the non-consumable wire can be used again and again. The workpiece is linked to the anode and the conductive wire is associated with the cathode. The wire μ-EDM is largely used in the manufacturing of micro-sized fins and screws. The machining of SiCp/Al-composite was conducted through wire μ-EDM for investigating MRR and SR. The rise in MRR was seen with an upsurge of wire tensional force and pulse-on time and a cut in the voltage and pulse-off time. The SR rose with voltage and pulse-on time but augmented initially and later decreased with an increase in the wire tensional force. A combination of optimum input parameters obtained through a multi-objective optimization approach raised the MRR from 12.6% to 49.1% and reduced the SR from 5.9% to 37.8% (Chen et al. 2021). The analysis of cutting speed, overcut, and MRR followed characterization was performed while fabricating the precise micro-fins of 850 μm in mean height. It was seen that all the responses amplified with a rise of peak current and pulse-on time. The optimum parameter combination for overall evaluation criteria was a pulse-on time of 5 μs, pulse-off time of 15 μs, peak current of 4 A, and table feed of 4 μm/s (Debnath & Patowari 2019).

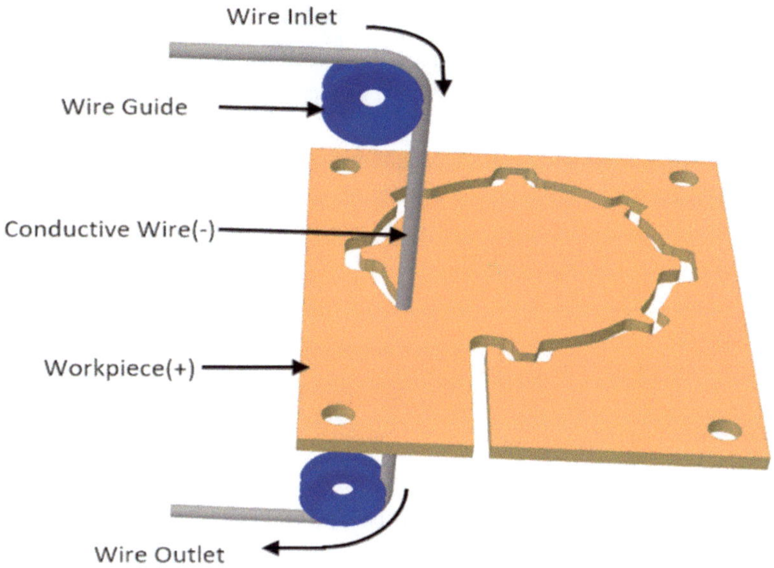

Fig. 6 Schematic representation of the wire μ-EDM process

3.6 Vibration-Assisted μ-EDM

In the modern world, high miniaturization demand requires continuous improvement in advanced machining technology. The μ-EDM is capable of manufacturing complex micro parts for various applications such as automobiles, nuclear, aerospace, etc. But when we must drill a deep-small diameter hole then it is very difficult to machine at that ratio. The main problem is the evacuation of the dielectric fluids along with the trash from the tiny area of the hole. To counter this issue, vibration in workpieces, tool electrodes and dielectric fluid was introduced. Figure 7 shows that vibrations are employed to the workpiece, and because of that, cavitation is formed, leading to the generation of negative pressure. Consequently, all the debris material is separated from the small area of the drill hole and stable machining is continued. Also, the machining performance is increased and the time required for machining is reduced. The imparting of ultrasound vibration resulted in adequate improvement of response features in the machining of a titanium alloy viz. 32% reduction in overcut, 73% decrease of TA, 65.8% reduction of TWR, and 2.4 time increase in MRR. The optimal processing parameter combination was established for each of the response features. The development of the predictive process model gave the MRR as 485,146.708 μm^3/s, TWR as 0.0292, TA as 0.05°, and overcut as 92.5232 μm. The results were found to be consistent with processed results and verified through tests, thereby proving the model's accuracy (Xing et al. 2020).

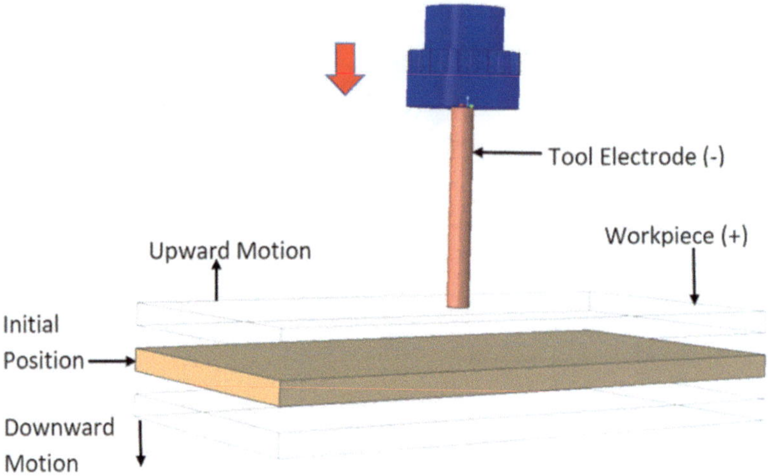

Fig. 7 Schematic representation of vibration-assisted μ-EDM process

4 Input Parameters of μ-EDM

The input process factors in μ-EDM are very significant from the machining point of view in terms of obtaining precise outcomes. It encompasses parameters both in relation to machining and machining setup. The machining parameters are again classified into two collections viz. non-electrical-based and electrical-based. The description of the input parameters is summarized below.

4.1 Discharge Energy

Discharge energy is arguably one of the parameters with utmost significance in μ-EDM. During the generation of each spark, the average electrical energy generated is termed 'discharge energy'. For the accomplishment of the machining, this energy gets converted into thermal energy (Gostimirovic et al. 2012). The discharge energy induced in the workpiece-tool gap in a pulse duration must be appreciably small to facilitate micromachining by elimination of materials in smaller units. It is represented in Eq. 1 (Liu et al. 2010).

$$\mathrm{D_e} = \int_0^{t_p} d_v(t).d_c(t)dt \tag{1}$$

where '$\mathrm{D_e}$' is the discharge energy in Joules, 't_p' is the duration of the pulse in seconds, 'd_v' denotes sparking voltage in volt, and 'd_c' denotes discharge current in amperes.

The factors responsible for controlling the discharge energy (D_e) are t_p and d_c, wherein a large value of the former and a small value of the latter are favourable for maintaining the discharge energy in an efficient way (Liu et al. 2010). The microhole quality can be determined by the kind of generator used in μ-EDM. This is mainly because the desired low value of discharge energy for a high surface finish varies from one kind to the other. The resistance–capacitance (R–C) generator produces lower discharge energy in comparison to the transistor kind merely by utilizing a small capacitance value (Jahan et al. 2009b). The discharge energy per unit pulse in the case of an R–C generator is given by Eq. 2 (Karthikeyan et al. 2012).

$$D_{e(R - C)} = \frac{1}{2}CV^2 \tag{2}$$

where 'V' and 'C' denote gap voltage and capacitance, respectively.

The capacitance decides both the pivotal factors, viz., duration of pulse and discharge current in the μ-EDM process. The sparks are discharged by the capacitor which stores in the form of charge. The point of current flow for ionising the plasma channel is governed by the voltage (Karthikeyan et al. 2012). The capacitance plays a much bigger role than the voltage in controlling the discharge energy and it has been unanimously agreed upon by numerous researchers (Jahan et al. 2009b). For some applications of μ-EDM, the transistor generator is used. The discharge energy per unit pulse in the case of a transistor generator is given by Eq. 3 (Son et al. 2007).

$$D_{e(T)} = V_{pulse} I_{pulse} \frac{t_{on}}{t_{on} + t_{off}} \tag{3}$$

here V_{pulse}, I_{pulse}, t_{on}, and t_{off} denote unit pulse voltage, unit pulse current, pulse-on time, and pulse-off time.

The term '$\frac{t_{on}}{t_{on}+t_{off}}$' is called the duty ratio, denoted by the term ψ. Therefore Eq. 3 can be written as shown in Eq. 4 (Jahan et al. 2009b).

$$D_{e(T)} = V_{pulse} I_{pulse} \psi \tag{4}$$

The duty ratio enacts an essential function in μ-EDM in which the transistor kind of generator is employed for micromachining operation (Jahan et al. 2009b). For achieving stability in the micromachining process, the values of t_{on} and t_{off} should be chosen suitably. It should be noted that the transistor kind of generator creates dimensions with low precision and inferior quality of surface in spite of increasing the efficiency of the process (Singh Kar et al. 2020a, b).

4.2 Polarity Effect

The designation of poles to both the electrode and workpiece is termed polarity. The straight polarity is attained when the workpiece is associated with the positive pole and the electrode is linked to the negative pole. On the contrary, reverse polarity occurs on interchanging the poles of connectivity of the electrode and the workpiece. The reverse polarity creates a high wear on the electrode (Kao, 2008). On the other hand, straight polarity is used in micromachining (Singh Kar et al. 2020a, b).

4.3 Tool Rotation

Tool rotation is essential in μ-EDM as it aids in flushing off debris from a miniature area by generating a centrifugal force. It is imperative to do so to escalate the rate of material removal from the workpiece (Pilligrin et al. 2017). Although conventional jet flushing is available for debris elimination, the method becomes unsuitable if the profile is too deep as the pressure of the jet becomes inadequate for assessing the location of inside contours. Also, an auxiliary arrangement with a vibrating holder for either the workpiece or tool can produce the desired stir of the micro debris. However, the setup is not desirable due to the complexity of the process. Consequently, the fluid has to be agitated by imparting a rotational motion to the tool for clearing off the micro-burrs in the micro-gap region (Karthikeyan et al. 2010).

4.4 Tool Feed

Traditional machining uses the term 'feed rate', implying the rate or speed at which the machining occurs. But the 'feed rate' in the context of the μ-EDM is not the same. The workpiece-tool gap must be maintained under all circumstances to accomplish the micromachining process. The rate of adjustment of the tool so that the requisite gap is properly sustained by the tool is called as feed rate. When the process experiences short-circuiting, a servo control aids in refeeding the tool after its retracting motion so that the desired gap is managed. This is when the feed rate comes into the picture. Also, whenever the gap becomes too large, or the sparks cease to occur, the feed rates regulate the self-positioning of tools with the adjoining workpiece point for the incidence of sparks. Also, unproductive machining time or idle time lessens with increasing the feed rate (Karthikeyan et al. 2010). But a high feed rate is not beneficial either, as the entrapped micro debris particles may initiate secondary sparks, thereby ruining the surface of the workpiece (Singh, Kar et al. 2020a, b).

4.5 Fixation of Tool and Workpiece

This is a non-electrical machining parameter that needs to be suitably controlled for achieving quality micro components in μ-EDM. The proper positioning of the tool and the workpiece is crucial for limiting errors and enhancing the dimensional accuracy of the machined parts. The shaft, hole, and end surface on the workpiece are taken as structures for positioning whereas the thin electrode is chosen as the tool. The hole positioning technique is particularly useful in μ-EDM with the wire variant. In this process, a hole is generated previously by drilling and the task is to insert the wire through it. The workpiece is positioned by the wire by trying to locate the central point of the hole drilled previously. The coordinate of this point is determined by travelling the wire along a two-dimensional axes system (X- and Y-coordinate) in successive moves. The positional error that is created should be minimal. It can be achieved by lowering the wire and hole diameter. A typical wheel or rod structure in the shape of a cylinder can be positioned with the help of the shaft positioning method. This method shares its similarity with the hole method with the exception of employing an electrode. After measuring and finding the results, the positioning of the shaft centre is accomplished. The end surface method for positioning is used for workpieces whose mutual surfaces are orthogonal to each other. The workpieces in the shape of a rectangle or square are suitable candidates for this method. The detection of the two orthogonal surfaces is done after properly clamping the workpiece, and thereafter, the positional error is determined by finding their ideal and true positions in a two-dimensional setup system (Wang et al. 2013). The tool experiences a positional error while mounted on the collet due to a lack of symmetry occurring on either side of the axis while machining, commonly referred to as tooltip shift. It can be decreased by ensuring that the tool is appropriately clamped. Also, another error in the form of tool breakage is generally seen in micromachining due to tool fixing and handling drawbacks (Karthikeyan et al. 2014). The clamping errors can be avoided in μ-EDM by fabricating the tool on the machine itself. The frequent change of microelectrodes induces re-setup or re-clamping of the same resulting in poor accuracy and precision of micro-components (Rahman et al. 2007).

4.6 Repositioning of the Tool

The ability to reposition a tool is essential for machining micro-slots with high complexity as it assists the workpiece in being positioned with good precision (Singh, Kar et al. 2020a, b). In μ-EDM, the inability to achieve repeatability and accuracy to position the micromachining setup leads to the generation of errors. The hole dimensions are greatly influenced by the repeatability of positioning whereas the hole position is affected by the accuracy of positioning (Pham et al. 2004).

5 Outputs in μ-EDM

There are a number of output parameters in μ-EDM which are responsible for assessing and typifying the micro features. The most important among these are the tool wear rate (TWR), material removal rate (MRR), taper angle (TA), recast layer thickness (RLT), surface roughness (SR) and overcut. The detailed explanation of these performance parameters is covered below.

5.1 *Material Removal Rate*

The removal of substance from workpieces during the μ-EDM process can be quantified in terms of volume for each time unit, which is termed the 'MRR'. Efficient micromachining is attained with a rise in MRR. The workpiece is weighed in the pre-machining stage (W_{pre}) and post-machining stage (W_{post}) and their numerical difference is calculated per unit of real machining duration (T_r) to get the value of MRR (Tiwary et al. 2014). The MRR can be expressed in a form as shown in Eq. 5.

$$\text{MRR} = \frac{W_{pre-W_{post}}}{T_r} \tag{5}$$

A rise in voltage decreases the availability of energy per unit area as the volume of the plasma expands and consequently lowers the MRR. The upsurge in capacitance plays a major role in MRR enhancement as it raises the energy in every single spark (Saxena et al. 2015).

5.2 *Tool Wear Rate*

Another essential response characteristic of the μ-EDM is the TWR defined as the amount of substance eliminated from the tool by erosion in volume per unit of real machining duration. The accuracy of the micromachining process is inversely proportional to the TWR (Tiwary et al. 2014).

5.3 *Taper Angle*

During the μ-EDM, the formed micro hole is tapered from the point of entry to the point of exit. The angle subtended by this slope with the vertical protruding from the exit point of the hole is known as the taper angle (TA) (Singh et al. 2020b). The TWR has a role to play in deciding the TA, as the geometrical features of the

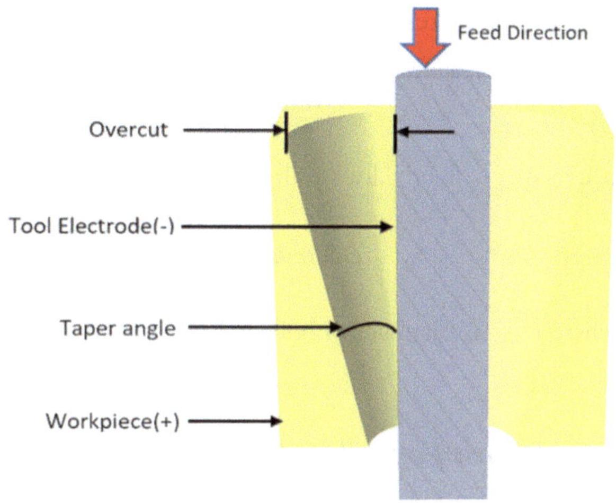

Fig. 8 Schematic representations of the taper angle and overcut

machined micro-hole depend on the physical structure of the tool. A high value of TA is observed for a straight polarity as compared to reverse polarity mainly because of the increasing TWR. A large tool rotational speed creates secondary discharge in the workpiece-tool gap and triggers volatility thereby increasing the TA (Pilligrin et al. 2017). The TA can be expressed by Eq. 6.

$$\mathrm{TA} = \tan^{-1}\left(\frac{Diameter_{(entry)} - Diameter_{(exit)}}{2k}\right) \tag{6}$$

where 'k' depicts the micro-hole depth. The TA is schematically illustrated in Fig. 8.

5.4 Overcut

Overcut is termed as one-half of the numerical subtraction in diameters of the machined micro-hole and tool. The discharge energy rises with current and eliminates material in the form of a large crater, and hence, more material elimination takes place than expected. The probability of gas bubble formation increases with high discharge energy and may flow away some of the micro debris along with it from the machined area, resulting in undesired secondary sparks. These conditions create some unsuspected removal of material heading towards a high overcut. It is observed from Eq. 2 that as the discharge energy directly varies with the gap voltage, so a high value of the latter will increase the overcut. Also, as the workpiece-tool gap increases and the discharges become less intense, the rate of material elimination reduces, and so does the overcut. Overcut is particularly beneficial too as it prevents short-circuiting by suitably managing the workpiece-tool gap (Singh et al. 2020a). The overcut is shown in Fig. 8. The overcut for micro slot and micro hole is shown by Eqs. 7 and 8 respectively.

$$\text{Overcut (micro slot)}=\frac{W_{ms}-D_{tool}}{2} \tag{7}$$

where W_{ms} is micro slot width and D_{tool} is tool diameter.

$$Overcut\ (micro\ hole)\ =\frac{D_{hole}-D_{tool}}{2} \tag{8}$$

where D_{hole} is diameter of hole and D_{tool} is tool diameter.

5.5 *Recast Layer Thickness*

The μ-EDM process generates cracks, protrusion, and other unwanted features on the surface post-machining, leading to recast layer development on its surface, as represented in Fig. 9. A few noteworthy studies have been done in order to upgrade the quality of microstructures by effective elimination of recast layers (Chu et al. 2020). The change induced regarding polarity is also found to be effective in trimming the layer to lower its thickness. The surface characteristics of the recast layer are also reliant on the use of dielectrics (Li et al. 2019). The recast layer thickness (RLT) needs analysis by building numerical models as the components are machined at the micro-scale to enable good control of results.

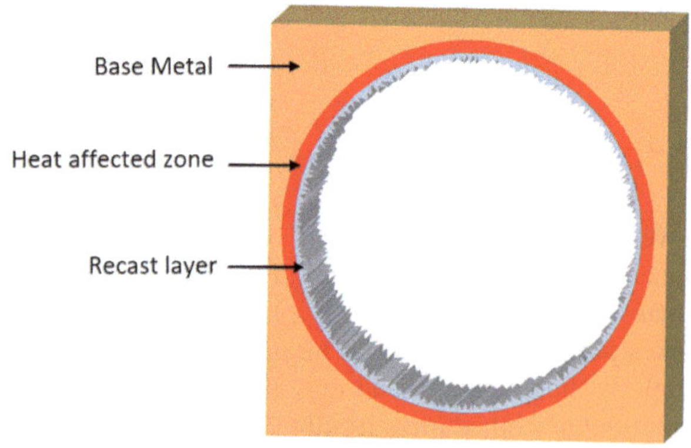

Fig. 9 Recast layer formation

6 Tool and Workpiece Materials

The energy of the pulse needs to be less to achieve a superfine finishing in μ-EDM. Moreover, the electro-thermal traits of tool electrodes display a substantial part in deciding the characteristics of a machined μ-EDM surface. The evaluation of appropriate tool materials for the μ-EDM is decided by the effect they exhibit on response characteristics like MRR and TWR. Tungsten carbide (WC) exhibits high hardness, along with some of its variants (WC–Co), designated as composites, particularly suited for cutting tools and die-over temperature ranges. Researchers are focussing on creating micro features with prime importance in relation to surface finish while machining WC parts. Many different electrodes are used in μ-EDM of WC. The typical electrodes include brass, silver, graphite, tungsten, and copper. Although copper (Cu) has been extensively used as an electrode material for μ-EDM, it may be unsuitable sometimes for the machining of materials like WC as it has low tool wear resistance (Zwilsky & Langer 1990). Some materials with steep melting points are used with Cu-based electrodes to overcome the above limitations. The suitable electrodes for μ-EDM of WC are found to be tungsten (W), copper-tungsten (Cu-W), and silver-tungsten (Ag-W). The surface properties are expected to vary with the implementation of diverse electrode tools. The alteration of the aforesaid properties is more pronounced when Cu-W and W electrodes are used than when Ag-W is used. Also, the crater height is low with the Ag-W electrode causing a small variation of peak to valley. In terms of MRR, it can be said that the Cu-W electrode is better than the Ag-W electrode owing to its greater melting point. But it cannot be opined with certainty that melting point is the deciding factor for choosing an electrode for MRR as the W (higher value of melting point) electrode gives a comparatively poorer MRR than the other two in μ-EDM finishing operations (Jahan et al. 2009a). The study also reveals the μ-EDM of ceramics viz. Si_3N_4-TiN with the help of a WC electrode. The highest (0.364 mm^3/min) and lowest (0.002 mm^3/min) values of MRR were obtained for roughing and finishing operations, respectively (Liu et al. 2010). Other materials that are often considered workpieces in μ-EDM are steel and silicon. The TWR for steel is higher than silicon for identical spark energy consideration (Song et al. 2000). The electrode wearing ratio in volume varies inversely with a few traits of the electrode material, viz., thermal conductivity, melting point and boiling point. The resistance to wear for a tungsten (W) electrode is the best in μ-EDM from an experimental point of view. Also, it is vital to take electrode rods with larger diameter to lessen the ratio of electrode wearing for any material (Tsai & Masuzawa 2004).

7 Circuits of the μ-EDM

The demand for the formation of surfaces with exquisite quality by the μ-EDM process requires enormously low material removal per discharge. This low value is again reliant on the energy of discharge, primarily managed by the pulse-on time and peak current. However, it is always preferred to reduce the pulse-on-time duration rather than reducing the peak current to get good outcomes in terms of surface quality. The resistance–capacitance (R–C) type of pulse generator is mostly used to satisfy the criteria of micromachining as it has the capability to produce a minute pulse-on-time duration in the range of nanoseconds. It may still lack efficient machining ability due to its inability to obtain substantial discharging frequency. There is another kind, commonly known as the transistor pulse generator, used very rarely in the perspective of the μ-EDM. The transistor generator produces a high discharging frequency and hence, induces a greater rate of material elimination. The transistor-type isopulse generator is a sort of transistor generator that can be successfully applied for μ-EDM applications (Han et al. 2007). A detailed discussion of the R–C generator and transistor generator is presented below.

7.1 *Resistance–Capacitance Generator*

A resistance–capacitance (R–C) circuit comprises a capacitance and a resistance. Whenever an electrical field is developed, the storing of electrical charges in the domain of the developed field is accomplished by the capacitor. An R–C circuit is shown in Fig. 10. The capacitor (C) and resistor (R) and are found in the circuit, and voltage is supplied in pulses by a source 'S'. The gap of the spark is modelled and comprises two parallelly connected components, viz., the gap of the spark and the equivalent capacitor in the gap C_d. The loop of discharging further comprises the resistance R_d and the inductor L_d. Also, the voltage across C is represented by s_v and the voltage across the spark gap is represented by s_{gap}. The circuit current is depicted by i_{Ld}. The value of s_v becomes negative as the initial normal discharge terminates along with a high value of s_{gap}. The ignition of the spark gap takes place because of the sparks owing to the significantly low time duration in μ-EDM (Yang et al. 2018). The discharging circuit explains the current oscillations in the μ-EDM. The stray inductance facilitates the current oscillations. During the discharging process, the plasma channel encounters a flow of current towards it. Kirchhoff's equations show the flow of current as given by Eqs. 9 and 10 (Cheong et al. 2018).

$$\frac{d^2}{dt^2}I + \frac{1}{R_d C_d}\frac{d}{dt}I + \frac{1}{L_d C_d}I - \frac{1}{L_d}\frac{d}{dt}s_v - \frac{1}{R_d L_d C_d}s_v = 0 \tag{9}$$

$$\frac{1}{C}I + \frac{d}{dt}s_v + \frac{1}{RC}s_v = \frac{1}{RC}S \tag{10}$$

where 'I' denotes the current flowing towards the tool from the workpiece.

Thereafter the charging of the capacitor (C) occurs with the subsequent dissipation of the plasma channel with termination of the current oscillations. The phenomenon of charging occurs as depicted in Eq. 11 (Cheong et al. 2018).

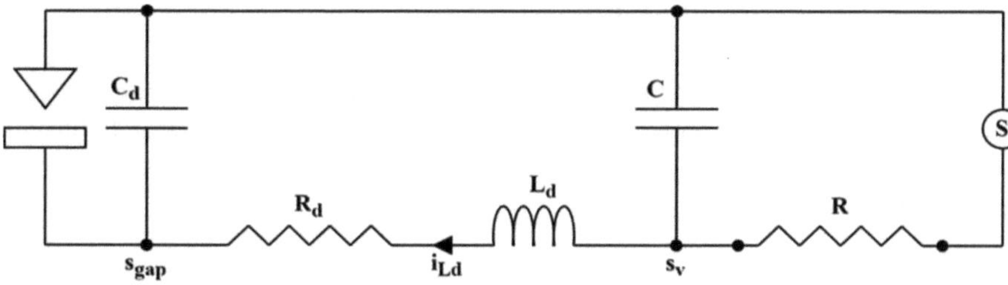

Fig. 10 R–C kind circuit (Yang et al. 2018)

$$s_v = S(1 - e^{\frac{-t}{RC}}) \tag{11}$$

As a consequence of the effect of the stray inductance, the direction of flow of current reverses i.e., it flows towards the workpiece from the tool (Cheong et al. 2018).

7.2 *Transistor Generator*

In a transistor pulse generator, the fundamental parameters like discharge time, discharge current and voltage influence the power of discharges (Chu et al. 2016). It is of paramount importance that the delay time be shortened while using an isopulse transistor generator in μ-EDM since the process demands isoduration discharging the pulse of current. The use of the pulse control circuit needs to be eliminated and the discharge current must shut off. The discharge circuit is devoid of any complicated integrated circuits, with just a combination of a few field effect transistors (FET). A study of the waveform reveals the attainment of around 30 ns pulse duration from the generated circuit (Han et al. 2004).

7.3 *Comparison of R–C Generator and Transistor Generator*

A comparative analysis of the R–C generator and transistor generator is elucidated by keeping in view several factors. The comparisons are made based on crater diameter, discharge waveforms, and radii of the discharge channel.

7.3.1 Based on Crater Diameter

On the comparison front of the two kinds of pulse generators, it is seen that the diameter-depth ratio of formed craters is greater for an R–C kind as compared to the transistor kind. This is because the density of energy and the obtained current are low for the transistor generator. The diameter and depth of the crater become wide and deep with the rising value of discharge voltage at a constant discharge capacitance for an R–C kind generator. For a transistor generator, a single pulse discharge creates a crater with substantial depth and width at a varying discharge time and constant discharge current because of the increase in energy and subsequent plasma channel expansion (Chu et al. 2016).

7.3.2 Based on Discharge Waveforms

In the circuit where discharge occurs, the stray capacitance is found to be prevalent. The effect is that the waveform of discharge for both the R–C and transistor kind of generators is analogous in the preliminary phase. The waveform of the R–C generator exhibits attenuation whereas the waveform of the transistor-kind generator achieves stability after a certain time duration (Chu et al. 2016).

7.3.3 Based on Radii of Discharge Channel

It is seen that for higher energy, a satisfactory growth of the plasma leads to a larger ultimate expansion radius for the transistor generator than the R–C generator. However, only a few differences in the ultimate expansion radius between the two kinds of generators are observed for low energy owing to the short discharge duration. The lengthier width of pulse at high energy in the case of the transistor kind of generator than the R–C kind of generator is the key to the generation of a greater radius of expansion (Chu et al. 2016).

8 Study of Dielectrics and Their Capabilities

Kerosene is one of the most popularly used dielectric fluids in μ-EDM. The pollution instigated by build-up of carbonaceous units of kerosene on surfaces prevents the stability and efficiency of sparks, reducing the MRR. Consequently, the utilization of desalinated water is chosen as dielectrics in μ-EDM. Moreover, the use of powders in various sizes and concentrations was done to investigate their effects on μ-EDM (Mohri & Saito 1991). The atmosphere cornering the machining with de-ionized water as dielectric promotes safety as no carbonaceous particles get deposited. It also improves some response parameters of μ-EDM like TWR and MRR. Ti6Al4V is an alloy of titanium machined by using kerosene mixed with powders of silicon carbide (SiC) and aluminium (Al) displaying different results of TWR and MRR depending on the energy of sparks and concentration of powders (Chow et al. 2000). The dielectric fluids are mixed with different metallic powders viz. copper (Cu), silicon (Si), chromium (Cr), aluminium (Al), etc. The addition of the C, Fe, Al, and Cu in kerosene enhanced its features related to breakdown. The addition of silicon in powdered form with dielectrics creates surfaces with glossy texture and generates low impact on workpieces eliminating the possibility of short-circuiting. A mirror-like finish of the surface is found by implementing dielectrics by mixing Al, Mo, and SiC with it. However, the finishing is highly dependent on the substances used as workpieces, the electrode polarity and the features of powders. The material removal is greatly influenced by heat conductivity, electrical resistivity, density, and dimension of the added powders in the dielectric medium. The reduction in TWR and escalation of MRR is solely dependent on adding the correct proportion of powder in the dielectric medium. Also, it can be stated that a low TWR and high MRR are obtained from small particle sizes while fixing the concentration of various particles added (Tzeng & Lee 2001). In the machining process, the generation of debris is taken care of by the proper flushing mechanism of dielectrics. The imparting of orbital movement to the workpiece and tool is also a feasible option for flushing (Wong et al. 1995).

9 Summary of Research Perspectives in μ-EDM

The diverse trait of the μ-EDM process is its applicability to use it in conjunction with processes like electrochemical machining, electrochemical etching, ultrasonic machining, and laser machining. The μ-EDM process can be used successfully in servo-controlled methods, ultrasound vibration-assisted methods, and nano powder-based addition methods in futuristic applications (Rajurkar 1994). A study evaluates the use of micromachining in sequences for the reduction of errors related to tool repositioning. This is vital from the point of view of lowering the production time, reducing scraps, and enhancing the level of accuracy (Chavoshi et al. 2017). The μ-EDM process can also be aided by several energy-assisting external means like plasma-aided, laser-aided, cryogenic-aided, and thermal-aided. The fabrication of micro rods with a substantially higher aspect ratio is possible by reverse μ-EDM process (Mujumdar et al. 2015).

10 Conclusion

The fascinating aspect of μ-EDM is the ease of producing micro-features on a large scale with high dimensional accuracy on the hard substances. This cutting-edge technique is even successful as a deposition method, coating tools, providing colourful films on materials, and fabricating sphere-like probes. The unwanted machining burrs can also be successfully removed by the μ-EDM. Researchers have further revealed the versatility of the process in machining non-conductive ceramic substances. However, some major issues cornering μ-EDM need to be addressed, viz. TWR should be controlled, dielectric flushing methods must be upgraded, and research on novel material for electrodes must be done, boosting the capability of various μ-EDM processes and producing sustainable μ-EDM. This chapter is an effort to bring out the underlying principles of μ-EDM, a discussion of its variants, an illustration of various parameters in μ-EDM, a comparative sketch of μ-EDM circuits, and the study of tool, workpiece, and dielectrics in μ-EDM. The prospects of μ-EDM in diverse areas of micro-manufacturing are also reviewed in brief.

References

Boral, S., Sidhu, S. S., Chatterjee, P., Chakraborty, S., & Gugaliya, A. (2019). Multi-response Optimization of Micro-EDM Processes: A State-of-the-Art Review. In *Micro-electrical Discharge Machining Processes, Materials Forming, Machining and Tribology* (pp. 293–310). Springer Singapore. https://doi.org/10.1007/978-981-13-3074-2

Chavoshi, S. Z., Goel, S., & Morantz, P. (2017). Current trends and future of sequential micro-machining processes on a single machine tool. *Materials & Design, 127*, 37–53. https://doi.org/10.1016/j.matdes.2017.04.057

Chen, Z., Zhou, H., Yan, Z., Han, F., & Yan, H. (2021). Machining characteristics of 65 vol .% SiCp / Al composite in micro-WEDM. *Ceramics International, 47*(10A), 13533–13543. https://doi.org/10.1016/j.ceramint.2021.01.212

Cheong, H., Kim, Y., & Chu, C. (2018). *Machining Characteristics of an RC-type Generator Circuit With an N-channel MOSFET in Micro EDM. 68*(April 2017), 631–636. https://doi.org/10.1016/j.procir.2017.12.127

Chow, H., Yan, B., Huang, F., & Hung, J. (2000). Study of added powder in kerosene for the micro-slit machining of titanium alloy using electro-discharge machining. *Journal of Materials Processing Technology, 101*(1–3), 95–103. https://doi.org/10.1016/S0924-0136(99)00458-6

Chu, X., Feng, W., Wang, C., & Hong, Y. (2016). Analysis of mechanism based on two types of pulse generators in micro-EDM using single pulse discharge. *The International Journal of Advanced Manufacturing Technology, 89*, 3217–3230. https://doi.org/10.1007/s00170-016-9287-8

Chu, X., Zhuang, W., Xue, W., Quan, X., Zhou, W., & Fu, T. (2020). Electrolytic removal of recast layers on micro-EDM microstructure surfaces. *The International Journal of Advanced Manufacturing Technology, 108*, 867–879. https://doi.org/10.1007/s00170-020-05410-x

Davim, J. P. (2013). *Nontraditional Machining Processes* (1st ed.). Springer London. https://doi.org/10.1007/978-1-4471-5179-1

Debnath, T., & Patowari, P. K. (2019). Fabrication of an array of micro-fins using Wire- EDM and its parametric analysis. *Materials and Manufacturing Processes, 34*(5), 580–589. https://doi.org/10.1080/10426914.2019.1566959

Gostimirovic, M., Kovac, P., Sekulic, M., & Skoric, B. (2012). Influence of discharge energy on machining characteristics in EDM †. *Journal of Mechanical Science and Technology, 26*(1), 173–179. https://doi.org/10.1007/s12206-011-0922-x

Han, F., Chen, L., Yu, D., & Zhou, X. (2007). Basic study on pulse generator for micro-EDM. *The International Journal of Advanced Manufacturing Technology, 33*, 474–479. https://doi.org/10.1007/s00170-006-0483-9

Han, F., Wachi, S., & Kunieda, M. (2004). *Improvement of machining characteristics of micro-EDM using transistor type isopulse generator and servo feed control. 28*, 378–385. https://doi.org/10.1016/j.precisioneng.2003.11.005

Jahan, M. P., Rahman, M., & Wong, Y. S. (2011). A review on the conventional and micro-electrodischarge machining of tungsten carbide. *International Journal of Machine Tools and Manufacture, 51*(12), 837–858. https://doi.org/10.1016/j.ijmachtools.2011.08.016

Jahan, M. P., Wong, Y. S., & Rahman, M. (2009a). A study on the fine-finish die-sinking micro-EDM of tungsten carbide using different electrode materials. *Journal of Materials Processing Technology, 9*, 3956–3967. https://doi.org/10.1016/j.jmatprotec.2008.09.015

Jahan, M. P., Wong, Y. S., & Rahman, M. (2009b). A study on the quality micro-hole machining of tungsten carbide by micro-EDM process using transistor and RC-type pulse generator. *Journal of Materials Processing Technology, 9*, 1706–1716. https://doi.org/10.1016/j.jmatprotec.2008.04.029

Kadirvel, A., Hariharan, P., & Mudhukrishnan, M. (2014). A study on the die-sinking micro-electrical discharge machining of EN-24 die steel using various electrode materials. *Advanced Materials Research, 984–985*, 73–82. https://doi.org/10.4028/www.scientific.net/AMR.984-985.73

Kao, C. C., & Ã, A. J. S. (2008). *Sub-nanosecond monitoring of micro-hole electrical discharge machining pulses and modeling of discharge ringing. 46*(2006), 1996–2008. https://doi.org/10.1016/j.ijmachtools.2006.01.008

Karthikeyan, G., Garg, A. K., Ramkumar, J., & Dhamodaran, S. (2012). A microscopic investigation of machining behavior in μED-milling process. *Journal of Manufacturing Processes, 14*(3), 297–306. https://doi.org/10.1016/j.jmapro.2012.01.003

Karthikeyan, G., Ramkumar, J., & Dhamodaran, S. (2014). BLOCK EDG: ISSUES AND APPLICABILITY IN MULTIPLE PASS μED-MILLING. *Machining Science and Technology, 18*(1), 37–41. https://doi.org/10.1080/10910344.2014.863650

Karthikeyan, G., Ramkumar, J., Dhamodaran, S., & Aravindan, S. (2010). Micro electric discharge milling process performance : An experimental investigation. *International Journal of Machine Tools and Manufacture, 50*(8), 718–727. https://doi.org/10.1016/j.ijmachtools.2010.04.007

Katz, Z., & Tibbles, C. J. (2005). Analysis of micro-scale EDM process. *Int J Adv Manuf Technol, 25*(2005), 923–928. https://doi.org/10.1007/s00170-003-2007-1

Kuriakose, S., Patowari, P. K., & Bhatt, J. (2020). Effect of micro-EDM machining parameters on the accuracy of micro hole drilling in Zr-based metallic glass. *Eng. Res. Express, 2*(015001). https://doi.org/10.1088/2631-8695/ab5c72

Li, C., Xu, X., Li, Y., Tong, H., Ding, S., Kong, Q., Zhao, L., Xu, X., Li, Y., Tong, H., Ding, S., Kong, Q., Zhao, L., & Ding, J. (2019). Effects of dielectric fluids on surface integrity for the recast layer in high speed EDM drilling of nickel alloy. *Journal of Alloys and Compounds, 783*, 95–102. https://doi.org/10.1016/j.jallcom.2018.12.283

Liu, K., Lauwers, B., & Reynaerts, D. (2010). Process capabilities of Micro-EDM and its applications. *International Journal of Advanced Manufacturing Technology, 47*(1–4), 11–19. https://doi.org/10.1007/s00170-009-2056-1

Marrocco, V., Modica, F., Bellantone, V., Medri, V., & Fassi, I. (2020). Pulse-Type Influence on the Micro-EDM Milling Machinability of Si3N4 – TiN Workpieces. *Micromachines, 11*(932). https://doi.org/10.3390/mi11100932

Mohri, N., & Saito, N. (1991). A New Process of Finish Machining on Free Surface by EDM Methods. *Annals of the CIRP, 40*(1), 207–210. https://doi.org/10.1016/S0007-8506(07)61969-6

Mujumdar, S. S., Mastud, S. A., Singh, R. K., & Joshi, S. S. (2015). Experimental characterization of the reverse micro-electrodischarge machining process for fabrication of high-aspect-ratio micro-rod arrays. *Proc. IMechE Vol. 224 Part B: J. Engineering Manufacture, 224*, 777–794. https://doi.org/10.1243/09544054JEM1745

Nikam, B., Shinde, B., Pandey, A., Pawade, R., Santana, A., Afonso, P., Zanin, A., & Wernke, R. (2018). Fabrication of Macro-Arrayed Structure using Reverse EDM: A Multi-objective Optimization. *Procedia Manufacturing, 20*, 160–167. https://doi.org/10.1016/j.promfg.2018.02.023

Pham, D. T., Dimov, S. S., Bigot, S., Ivanov, A., & Popov, K. (2004). Micro-EDM - Recent developments and research issues. *Journal of Materials Processing Technology, 149*(1–3), 50–57. https://doi.org/10.1016/j.jmatprotec.2004.02.008

Pilligrin, J. C., Asokan, P., Jerald, J., Kanagaraj, G., Nilakantan, J. M., Pilligrin, J. C., Asokan, P., Jerald, J., Kanagaraj, G., & Nilakantan, J. M. (2017). Tool speed and polarity effects in micro-EDM drilling of 316L stainless steel. *Production & Manufacturing Research, 5*(1), 99–117. https://doi.org/10.1080/21693277.2017.1357055

Rahman, M., Lim, H. S., Neo, K. S., Kumar, A. S., Wong, Y. S., & Li, X. P. (2007). Tool-based nanofinishing and micromachining. *Journal of Materials Processing Technology, 185*(1–3), 2–16. https://doi.org/10.1016/j.jmatprotec.2006.03.121

Raju, L., & Hiremath, S. S. (2016). A State-of-the-art Review on Micro Electro-Discharge Machining. *Procedia Technology, 25*(RAEREST 2016), 1281–1288. https://doi.org/10.1016/j.protcy.2016.08.222

Rajurkar, K. P. (1994). *Nontraditional Manufacturing Processes.Handbook of Design, Manufacturing and Automation.* John Wiley & Sons, Inc. https://doi.org/10.1002/9780470172452

Saxena, K. K., Srivastava, A. S., & Agarwal, S. (2015). Experimental investigation into the micro-EDM characteristics of conductive SiC. *Ceramics International, 42*(1), 1597–1610. https://doi.org/10.1016/j.ceramint.2015.09.111

Singh, A. K., Kar, S., & Patowari, P. K. (2020). Accuracy Improvement and Precision Measurement on Micro-EDM. In *Lecture Notes in Mechanical Engineering* (pp. 47–77). https://doi.org/10.1007/978-981-15-2117-1_3

Singh, A. K., Patowari, P. K., & Chandrasekaran, M. (2020). Experimental study on drilling micro-hole through micro-EDM and optimization of multiple performance characteristics. *Journal of the Brazilian Society of Mechanical Sciences and Engineering, 42*(10). https://doi.org/10.1007/s40430-020-02595-w

Singh, A. K., Patowari, P. K., & Deshpande, N. V. (2019). Analysis of micro-rods machined using reverse micro-EDM. *Journal of the Brazilian Society of Mechanical Sciences and Engineering, 41*(1). https://doi.org/10.1007/s40430-018-1519-4

Son, S., Lim, H., Kumar, A. S., & Rahman, M. (2007). Influences of pulsed power condition on the machining properties in micro EDM. *Journal of Materials Processing Technology, 190*, 73–76. https://doi.org/10.1016/j.jmatprotec.2007.03.108

Song, X., Meeusen, W., Reynaerts, D., & Brussel, H. Van. (2000). Experimental Study of Micro-EDM Machining Performances on Silicon Wafer. *In Micromachining and Microfabrication Process Technology VI,Proceedings of SPIE, 4174*, 331–339. https://doi.org/10.1117/12.396450

Tiwary, A. P., Pradhan, B. B., & Bhattacharyya, B. (2014). Application of multi-criteria decision making methods for selection of micro-EDM process parameters. *Advances in Manufacturing, 2*, 251–258. https://doi.org/10.1007/s40436-013-0050-1

Tsai, Y., & Masuzawa, T. (2004). An index to evaluate the wear resistance of the electrode in micro-EDM. *Journal of Materials Processing Technology, 149*(1), 304–309. https://doi.org/10.1016/j.jmatprotec.2004.02.043

Tzeng, Y., & Lee, C. (2001). Effects of Powder Characteristics on Electrodischarge Machining Efficiency. *The International Journal of Advanced Manufacturing Technology, 17*, 586–592. https://doi.org/10.1007/s001700170142

Uhlmann, E., Piltz, S., & Doll, U. (2005). Machining of micro/miniature dies and moulds by electrical discharge machining - Recent development. *Journal of Materials Processing Technology, 167*(2–3), 488–493. https://doi.org/10.1016/j.jmatprotec.2005.06.013

Urso, G. D., Giardini, C., & Quarto, M. (2018). Characterization of surfaces obtained by micro-EDM milling on steel and ceramic components. *The International Journal of Advanced Manufacturing Technology, 97*, 2077–2085. https://doi.org/10.1007/s00170-018-1962-5

Vidya, S., Wattal, R., & Rao, P. V. (2021). Investigation of machining performance in die-sinking electrical discharge machining of pentagonal micro-cavities using cylindrical electrode. *Journal of the Brazilian Society of Mechanical Sciences and Engineering, 43*(6), 1–9. https://doi.org/10.1007/s40430-021-03012-6

Wang, C., Chu, X., Liu, G., Zhu, K., Hong, Y., & Chi, G. (2013). The Design of Integrated Route in Micro-EDM. *Materials and Manufacturing Processes, 28*(12), 1348–1355. https://doi.org/10.1080/10426914.2013.832301

Wong, Y. S., Lim, L. C., & Lee, L. C. (1995). Effects of Flushing on Electro-Discharge Machined Surfaces. *Journal of Materials Processing Technology, 48*(1–4), 299–305. https://doi.org/10.1016/0924-0136(94)01662-K

Xing, Q., Yao, Z., & Zhang, Q. (2020). Effects of processing parameters on processing performances of ultrasonic vibration-assisted micro-EDM. *The International Journal of Advanced Manufacturing Technology*, *112*, 71–86. https://doi.org/10.1007/s00170-020-06357-9

Yang, F., Qian, J., Wang, J., & Reynaerts, D. (2018). Simulation and experimental analysis of alternating-current phenomenon in micro-EDM with a RC-type generator. *Journal of Materials Processing Technology*, *255*, 865–875. https://doi.org/10.1016/j.jmatprotec.2018.01.031

Zwilsky, K. M., & Langer, E. L. (1990). *VOLUME 2 Properties and Selection:Nonferrous Alloys and Special-Purpose Materials*. ASM International. https://doi.org/10.31399/asm.hb.v02.978 1627081627

Studies and Comparison of Red Mud and Neutralised Red Mud Using Waste Lemon Biomass Juice for Composite Reinforcement Application

Priyeshiv Kumar Gurmaita, Rosang Pongen, and Shival Kumar Gurmaita

Abstract Red mud is a byproduct generated during the production of alumina in the aluminium industry that causes significant environmental and health problems. Red mud's powerful alkalinity has a significant impact on both its potential for use in composite reinforcement and its wide range of other applications. For the neutralisation of red mud, the method using natural waste lemon biomass and no chemical addition was suggested. Lemon juice was used to neutralise red muck, and the stirring process was studied. GC–MS can detect the presence of organic acids in lemon juice. Under experimental conditions, red mud's alkalinity can be effectively lowered by adding lemon juice, which also lowers the material's pH from 10.4 to 7.56. Numerous characterisations, including XRD, FTIR and TGA analyses, were conducted to study the neutralisation mechanism. The findings indicate that red mud undergoes minimal changes post-neutralisation, suggesting that small-molecule organic acids primarily eliminate free alkali without significantly altering the red mud itself without changing the chemical composition. In order to improve corrosion resistance, neutralised red mud was employed as a reinforcement in metal matrix composites.

Keywords Red mud · Neutralised red mud · Composite reinforcement application · XRD · FTIR · TG

P. K. Gurmaita (✉) · R. Pongen
Department of Mechanical Engineering, National Institute of Technology, Dimapur, Nagaland, India
e-mail: priyeshivkumarg@gmail.com

S. K. Gurmaita
Department of Civil Engineering, National Institute of Technology, Dimapur, Nagaland, India

M. Ram et al. (eds.), *Technological Advancements in Engineering and Management Sciences*, Management and Industrial Engineering,
https://doi.org/10.1007/978-3-032-04606-2_9

1 Introduction

Red mud, or bauxite residue, is a waste material generated during the process of refining bauxite ore to obtain aluminium oxide, the precursor for producing aluminium metal. Red mud is generated in large quantities, and managing its disposal poses a significant environmental challenge. However, in recent years, researchers have been exploring the application of red mud as a reinforcing material. Due to its availability, low cost, and potential geotechnical properties. Red mud is primarily composed of oxides of iron (Fe_2O_3), aluminium (Al_2O_3), and silicon (SiO_2), as well as other trace elements. Additionally, red mud can be stabilised by adding other materials, such as cement, to improve its strength and durability (Deelwal et al. 2014). It has been the subject of research for its utilisation in various industrial applications due to its abundant availability and environmental concerns associated with its disposal. Microstructure studies are conducted to understand the structure and composition of the cement mortar at a microscopic level. Techniques like XRD, SEM, and EDS are frequently employed to examine the microstructure of red mud. These analyses offer insights into the size, distribution, and composition of cement, red mud, and other components within the mortar (Anirudh et al. 2020). Characterising red mud is crucial for gaining insight into its properties, exploring possible applications, and evaluating its environmental implications. Ongoing research in materials science and environmental engineering focuses on developing innovative approaches for the safe and sustainable use of red mud (Nath et al. n.d). Al7075 alloy is recognised for its excellent mechanical characteristics, including high strength and corrosion resistance, making it an ideal choice for a matrix material. Incorporating Mica, Fly Ash, and Red Mud particles into the Al7075 matrix leads to the development of a hybrid composite, which exhibits improved wear resistance compared to the pure Al7075 matrix. This improved wear resistance is attributed to the abrasive and hardness properties of the reinforcement particles, which resist wear during sliding. The wear performance of the hybrid composite is significantly influenced by the size, shape, and distribution of the reinforcing particles (Matti Jan. 2021). It is a highly alkaline and toxic waste material, containing heavy metals, radioactive elements, and other contaminants. Despite its hazardous nature, red mud has been shown to have some useful properties, such as its high iron oxide content, which makes it a valuable source of iron and other metals. In India, red mud is generated in large quantities, and its proper utilisation and disposal are significant environmental challenges (Samal et al. 2013).The physical characteristics of red mud depend on its chemical makeup and the specific conditions under which it is produced. Key physical attributes include particle size distribution, bulk density, porosity, and surface area. These properties can affect the handling and storage of red mud, as well as its potential for reuse or disposal (Antunes et al. 2012). Red mud is a complex mixture of iron and aluminium oxides, silicates, and various impurities. The synthesis of nano-red mud involves various techniques such as chemical precipitation, hydrothermal synthesis, and the sol–gel method. The selection of a synthesis method is guided by the targeted size, shape, and purity of the nanoparticles. The characterisation of nano-sized red mud generally employs advanced analytical techniques such as XRD, TEM, thermogravimetric analysis (TGA), and FTIR. These methods collectively offer insights into the material's crystallographic structure, particle morphology, dimensional scale, and elemental or functional composition. Additionally, the thermal behaviour of red mud can be assessed using thermal analysis techniques like TGA and differential thermal analysis (DTA), which help determine its stability, decomposition profile, and phase transitions under controlled temperature conditions. These methods provide insights into the material's weight loss and temperature variations during heating. The data obtained can help identify the composition and characteristics of red mud, including any organic or inorganic substances that influence its thermal behaviour. Multifunctional mixed metal-biochar composites (MMBC) are advanced materials that merge the advantages of metal-based components with biochar. They are commonly produced using red mud, a waste from aluminium manufacturing, and lignin, a by-product from the pulp and paper industry. MMBCs have a range of applications in the fields of fabrication and the environment. In fabrication, MMBCs can be used as catalysts, supports for catalysts, and as active materials in energy storage and conversion devices. The metal-based components of MMBCs impart desirable properties such as high conductivity, thermal stability, and catalytic activity. The biochar component provides benefits such as high surface area, stability, and hydrophobicity (Cho et al. 2019). The active dealkalization process plays a key role in minimising the environmental impact of red mud, making it a safer and more sustainable material for aluminium production. When red mud is added as a reinforcement to Aluminium 6061 composites, it is expected to enhance certain mechanical and tribological properties. While tensile strength may decrease, improvements in wear resistance, hardness, and toughness are anticipated. This composite material has potential applications across various industries, including aerospace, marine, bicycle manufacturing, camera lenses, drive shafts, electrical components, brakes, and fittings. Utilising red mud as reinforcement also provides a valuable way to repurpose an industrial by-product from aluminium production. The current study focuses on neutralising red mud using lemon solution (SOL) to achieve an effectively neutralised material. Both untreated and neutralised red mud samples are analysed using XRD, FTIR, and TGA to enable the comprehensive utilisation of red mud as reinforcement in metal matrix composites.

2 Material and Experimental Methods

2.1 Material

Red mud samples were taken in NALCO, Odisha, India. Raw red mud was ball milled, sieved, and heated to 90 °C for 6 h before drying. In several Indian cities, including Dimapur, Nagaland, lemon trees are a common sight. Each year, there is a significant amount of waste produced. The raw materials for the waste lemon biomass were readily available waste lemons from lemon trees. In order to remove surface debris and pollutants, the samples were washed with water. After being crushed with a juicer, a lemon solution (SOL) with a pH of 4 was then created.

2.2 Experimental Methods

When red mud samples were mixed with water and stirred, electrical conductivity measurements indicated the presence of soluble components, primarily free alkali and salts. The red mud neutralised with lemon solution (SOL) was labelled RM/SOL, while the untreated red mud was referred to as RM. Following extraction with dichloromethane, the components were analysed using gas chromatography-mass spectrometry (GC–MS).

The red mud neutralization process using the solution of lemon (SOL) acid involves the following steps:

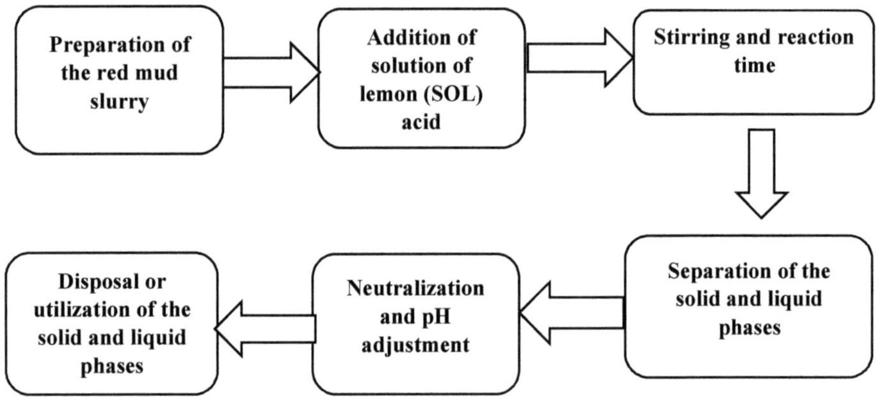

2.3 Characterization

A comprehensive set of characterisation techniques was employed to investigate the neutralisation process of red mud. Both untreated and neutralised red mud samples were analysed using TGA, FTIR, and XRD. Following complete drying, the red mud was finely pulverized with an agate mortar to obtain particles smaller than 65 μm. XRD analysis was utilised to determine the crystalline phases present in the samples, whereas FTIR spectroscopy was used to examine the functional groups and bonding characteristics. Additionally, TGA offered insights into the thermal behaviour of red mud by monitoring changes in mass with temperature, which indirectly revealed information about its chemical composition.

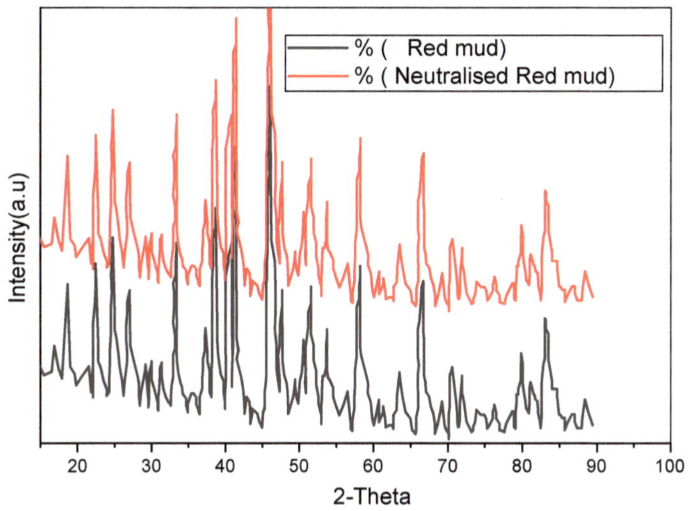

Fig. 1 XRD diagram of red mud and neutralised red mud

3 Results and Discussions

3.1 *XRD Analysis*

Red mud is composed of a variety of oxides and minerals, including iron oxide, aluminium oxide, silicon dioxide, sodium oxide, calcium carbonate, as well as mineral phases such as cancrinite, sodalite, and perovskite. Its overall chemical and mineralogical makeup can differ significantly depending on the bauxite ore source and the processing techniques employed during alumina extraction. XRD analysis was utilised to identify the crystalline phases present in the red mud sample, as depicted in Fig. 1. The XRD pattern revealed the presence of multiple mineral constituents, notably katoite, cancrinite, calcite, perovskite, hematite, silica, and brookite, confirming the heterogeneous nature of the red mud. Similar reports have been made before (Zhu et al. 2015). No noticeable changes in XRD patterns were seen when red mud was neutralised with SOL, suggesting that neutralisation may not have destroyed the red mud's main chemical content. Chemical content for metal matrix composite making may be present in the chemical composition.

3.2 *FTIR Analysis*

In order to determine the surface functional groups that may change during neutralisation, FTIR spectra were taken from organic acids in SOL. According to Fig. 2, the usual red mud FTIR peak locations are at 3550 cm^{-1}, 1450 cm^{-1}, and 1050

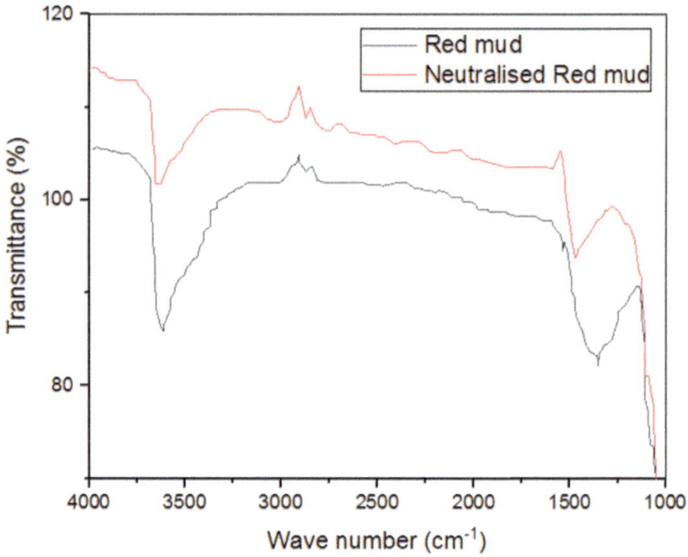

Fig. 2 FTIR diagram of red mud and neutralised red mud

cm^{-1}, respectively. These peak locations correspond to the -OH bond of water molecules, Si-O-Al, and Si-O bonds in red mud minerals. Peak displacement or the emergence of new peaks is not caused by neutralising red dirt with SOL derived from it. Further confirmation of this finding comes from FTIR analysis, which shows that neutralisation barely modifies red mud.

3.3 TGA Analysis

The TGA curves of red mud, both prior to and following neutralization, exhibit varying thermal behaviour rather than a uniform pattern. A noticeable weight reduction is observed across the temperature range of 150–800 °C. The initial phase of weight loss, marked by a peak near 250 °C, is associated with the release of physically bound water, which plays a critical role in the crystallization of red mud minerals. A subsequent stage of mass loss is observed between 500 °C and 700 °C, reaching its maximum around 600 °C, and is primarily linked to the decomposition of carbonate-based phases. The thermal response captured through TG analysis further substantiates the changes in red mud composition induced by the neutralization process (Fig. 3).

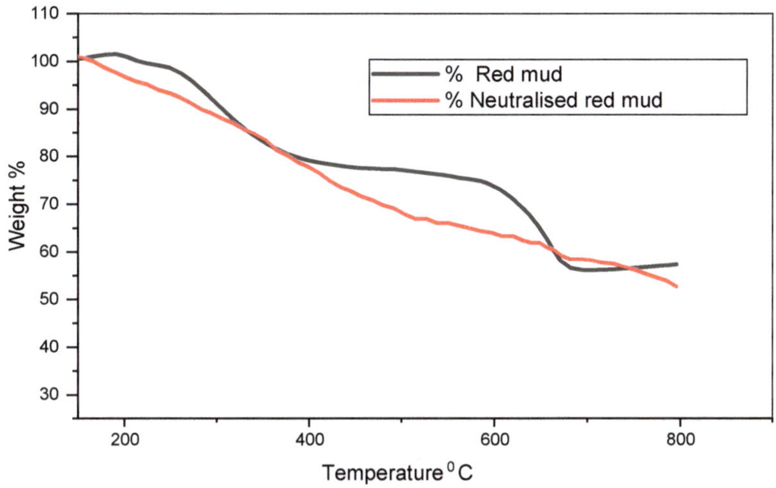

Fig. 3 TGA Curve red mud and neutralised red mud

4 Conclusions

- It was observed that the pH of the red mud decreased after neutralisation with a solution of lemon (SOL).
- According to XRD data, the red mud primarily contained the mineral types calcite, perovskite, illite, hematite, and magnetite.
- No peak shift or emergence of new peaks results from neutralisation of red mud with SOL produced from various chemical activation rates. Further confirmation of this finding comes from FTIR analysis, which shows that neutralisation barely modifies red mud.
- According to TGA findings, red mud's weight loss occurred at a higher temperature of 550 °C than red mud that had been neutralised, which was 600 °C. This suggests that red mud's thermal stability has risen and that, as a result, its reactivity may be reduced with longer storage periods.
- The paper demonstrates the feasibility of utilising the waste from the aluminium industry and biomass for the reinforcement of metal matrix composites (RAECS 2014).

References

M. Anirudh, K. S. Rekha, C. Venkatesh, and R. Nerella, "Characterization of red mud based cement mortar; Mechanical and microstructure studies," in *Materials Today: Proceedings*, 2020, vol. 43, pp. 1587–1591. https://doi.org/10.1016/j.matpr.2020.09.504.

Annual IEEE Computer Conference, International Conference on Recent Advances in Engineering and Computational Sciences 2014.03.06–08 Chandigarh, and RAECS 2014.03.06–08 Chandigarh, *Recent advances in engineering and computational sciences (RAECS), 2014 [international conference] ; 6–8 March 2014, University Institute of Engineering and Technology, Panjab University, Chandigarh, India.*

M. L. P. Antunes *et al.*, "Red mud from Brazil: Thermal behavior and physical properties," in *Industrial and Engineering Chemistry Research*, Jan. 2012, vol. 51, no. 2, pp. 775–779. https://doi.org/10.1021/ie201700k.

D. W. Cho *et al.*, "Fabrication and environmental applications of multifunctional mixed metal-biochar composites (MMBC) from red mud and lignin wastes," *J Hazard Mater*, vol. 374, pp. 412–419, Jul. 2019, https://doi.org/10.1016/j.jhazma t.2019.04.071.

K. Deelwal, K. Dharavath, M. Kulshreshtha, and A. Professor, "EVALUATION OF CHARACTERISTIC PROPERTIES OF RED MUD FOR POSSIBLE USE AS A GEOTECHNICAL MATERIAL IN CIVIL CONSTRUCTION," 2014.

Y. Feng and C. Yang, "Analysis on Physical and Mechanical Properties of Red Mud Materials and Stockpile Stability after Dilatation," *Advances in Materials Science and Engineering*, vol. 2018, 2018, https://doi.org/10.1155/2018/8784232.

S. Matti, "Dry sliding wear behavior of Mica, Fly Ash and Red Mud particles reinforced Al7075 alloy hybrid metal matrix composites," *Indian J Sci Technol*, vol. 14, no. 4, pp. 310–318, Jan. 2021, https://doi.org/10.17485/IJST/v14i4.2081.

H. Nath and A. Sahoo, "A STUDY ON THE CHARACTERIZATION OF RED MUD."

S. Samal, A. K. Ray, and A. Bandopadhyay, "Proposal for resources, utilization and processes of red mud in India — A review," *Int J Miner Process*, vol. 118, pp. 43–55, Jan. 2013, https://doi.org/10.1016/j.minpro.2012.11.001.

X. Zhu, W. Li, and X. Guan, "An active dealkalization of red mud with roasting and water leaching," *J Hazard Mater*, vol. 286, pp. 85–91, Apr. 2015, https://doi.org/10.1016/j.jhazmat.2014.12.048.

Assessment of Composite Insulated Rooftops' Energy Efficiency and Cost-Effectiveness for Sustainable Buildings in Indian Climate Conditions

Ruchita Shrimali, Naveen Kumar Agarwal, and Laxman Kumar Pandey

Abstract Buildings experience significant temperature fluctuations due to heat transfer through their external surfaces, with the roof typically being the largest contributor to heat intrusion. Incorporating high-quality insulation into the building envelope is essential for controlling indoor climate, cutting down on energy demands, and reducing the environmental impact of the built environment. In response to this, our study investigates how different insulation materials and their thicknesses influence a building's energy performance. The study focuses on eight cities across India, each representing different climatic conditions, to ensure the method is broadly applicable. By analyzing a composite insulation layer created from seven distinct insulation materials, we assess its performance using ANSYS software and e-Quest building energy modeling. Annual simulation results highlight that strategic insulation optimization greatly lowers energy consumption and associated CO_2 emissions. Energy demand reductions observed across different cases range from 30% to as high as 90%. Additionally, the optimized insulation strategies can lower indoor temperatures by up to 19 °C and cut greenhouse gas emissions by 90–96%. This strategy improves building sustainability while also contributing to a low-carbon future by reducing the environmental effects of energy use.

Keywords Thermal performance · insulation · energy modeling · optimum thickness · cooling loa · payback period

---

R. Shrimali (✉) · L. K. Pandey
Mechanical Engineering Department, A.C. Patil College of Engineering, Mumbai, Maharashtra, India
e-mail: shrimaliruchita1306@gmail.com

L. K. Pandey
e-mail: lkpandey@acpce.ac.in

N. K. Agarwal
Mechanical Engineering Department, Sir Padampat Singhania University, Udaipur, Rajasthan, India
e-mail: naveen.kumar@spsu.ac.in

M. Ram et al. (eds.), *Technological Advancements in Engineering and Management Sciences*, Management and Industrial Engineering,
https://doi.org/10.1007/978-3-032-04606-2_10

Nomenclature

A year, C	Yearly annual energy cooling cost (Rs/m^2 year)
CA,C	Annual cost of Energy (Rs/ m^2year)
CC	Annual energy cost per unit area of roof without insulation (RS/ m^2 year)
CDD	Cooling degree days
DD	Degree days
Ce	Unit cost of electricity (Rs)
C ins	Total insulation cost (Rs/m^2)
COP	Coefficient of Performance
Ct, c	Total cooling cost of an insulated building (Rs/m^2)
Cy	Cost of Insulation layer (Rs/m^3)
g	The inflation rate (%)
i	Rate of interest (%)
K	Heat transfer coefficient of the insulation materials (W/mK)
L	Thermal barrier (Insulation) thickness (m)
N	Lifetime (Years)
PPC	Payback Period for cooling (Years)
PWF	Present Worth Factor
r	Actual interest rate (%)
Rth	Total thermal resistance (m^2K/W)
U	Overall Hear transfer coefficient (W/m^2K)
ΔT	Difference between the outside ambient temperature and the constant inside ambient temperature (K)
x	Insulation thickness (m)
Xopt.,C	Optimum insulation thickness (cooling) (m)
EPS	Expanded Polystyrene
PUF	Polyurethane Foam

1 Introduction

The growing global demand for energy, driven by widespread air conditioning use in buildings, raises concerns about heat gain, particularly from roofs, which cover 20–25% of urban surfaces (González-Torres et al., 2022). By minimizing heat gains, buildings can reduce their energy requirements (Anand et al., 2015; Simona et al., 2017). Insulated roofs offer a sustainable solution to improve thermal comfort and reduce energy consumption (Aktemur et al., 2021; González-Torres et al., 2022). Rapid urbanization has made the construction industry a key driver of carbon emissions (Halwatura & Jayasinghe, 2008). Choosing the right insulation materials is crucial for reducing energy use and carbon footprints by enhancing thermal efficiency (Abdelzaher, 2023). While quality insulation may raise initial costs, it leads to long-term savings by cutting heating and cooling expenses (Mahlia et al., 2007; Mirrahimi et al., 2016). Building design, thermal mass, and climate also impact energy demand, and effective insulation helps lower both annual cooling loads and peak energy consumption (Kaynakli, 2008). Insulation materials function by minimizing heat flow through the three main processes: thermal radiation, direct conduction, and air movement or convection. These substances are specifically designed at a microscopic scale to block the movement of heat, slowing down thermal flow. When it comes to radiative heat, they work by reducing the release of energy as electromagnetic radiation. For conductive heat transfer, these materials slow the process by disrupting the direct transfer of energy between neighboring particles. For convection, these materials slow down the movement of air or liquids that can carry heat. Their structure is designed in a way that greatly reduces heat transfer, helping improve energy efficiency in many uses (Kumar & Suman, 2013; Friess & Rakhshan, 2017). Utilizing materials that block heat flow helps maintain a stable temperature by stopping heat from entering or leaving. This means less need for heating and cooling, which saves energy, lowers costs, and makes indoor spaces more comfortable (Ran et al., 2017; Wang et al., 2023). Dong et al. (2023) used DeST software to simulate building energy use and assess the impact of insulation materials on energy load and efficiency. Their research emphasizes how insulation contributes to better energy efficiency.

Although the roof contributes to 40–45% of heat loss or gain (Pearlmutter & Rosenfeld, 2008), a comprehensive study on roof insulation is still lacking in the literature. Given that the roof covers a larger, uniform surface exposed to outside air, and it's easier to apply thicker insulation to the roof than the walls, insulating the roof is generally more effective in terms of performance. Utilizing roof insulation stands out as a highly efficient method for curbing energy consumption in buildings (Boixo et al., 2012; Raut et al., 2019). Adopting cutting-edge building techniques, materials, and technologies is crucial for fostering environmental sustainability. This approach reduces heat loss and gain, contributing to energy efficiency, while also encouraging environmentally responsible construction practices. In particular, roof insulation plays a crucial role in enhancing building sustainability by improving energy efficiency, conserving natural resources, reducing environmental impact, and ensuring compliance with regulatory standards.

1.1 Framework

1.2 Comparative Study of Insulation Materials for Indian Climatic Conditions

India's diverse climatic zones, including hot and dry, warm and humid, composite, cold, and temperate regions (as illustrated in Fig. 1 for chosen 8 studies under study), have been the focus of numerous studies aimed at identifying optimal insulation thicknesses (Mishra et al., 2012; Mahapatra, 2022; Singh et al., 2016). India's hot and humid climate, particularly during summer, results from its tropical location near the equator, where intense sunlight increases temperatures and solar radiation. Urban buildings often consume large amounts of energy, which results in considerable greenhouse gas emissions and accelerates global warming. Implementing sustainable construction methods, especially with the use of effective insulation, can lead to a reduction in energy consumption by up to 77% and help lower carbon emissions. In alignment with this goal, India's National Mission for Sustainable Habitat emphasizes reducing electricity consumption as a primary objective. Given India's diverse climatic zones, the effectiveness of insulation varies. Insulation materials regulate temperature, conserve energy, and improve comfort. Table 2 compares key materials—EPS, Fiberglass, Foam Concrete, Peripor, Neopor, and Polyurethane Foam—based on their properties (Karpov et al., 2021; Jalal & Tanveer, 1991; Rajak et al., 2021), while Table 1 provides thermophysical values used for energy modeling in e-Quest software.

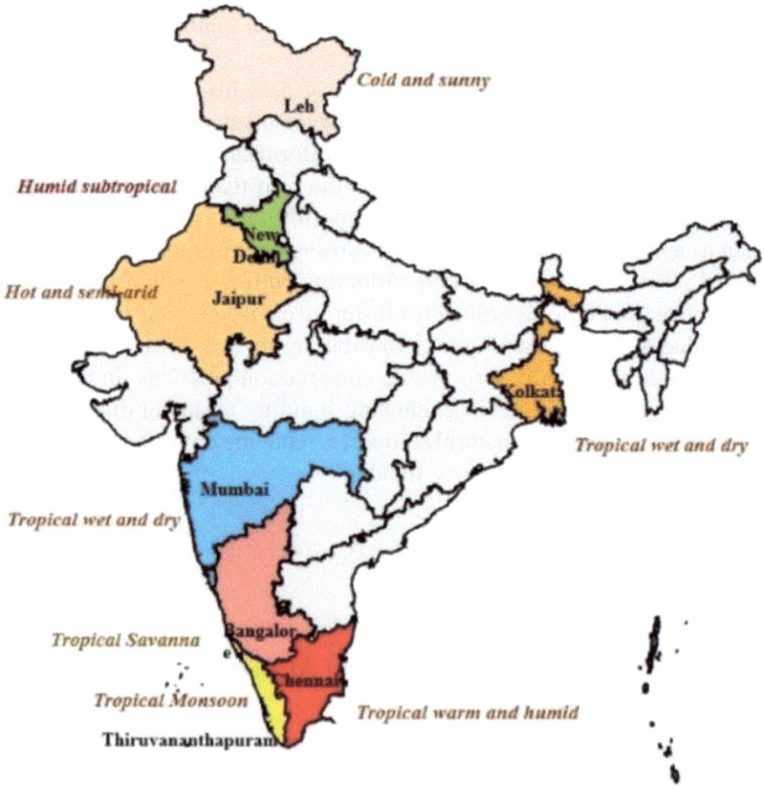

Fig. 1 Geographical map of India highlighting eight selected cities from various climatic zones (Shrimali et al., 2024)

Table 1 Thermophysical properties of chosen insulation materials (Shrimali & Agrawal, 2024)

Insulation material	Density kg/m^3	Specific heat J/Kg °C	Thermal transmission W/ m–K	Thermal resistance m^{2}K/W
Expanded (foam) Polystyrene (EPS)	Nov-32	1300–1500	3.8×10^{-2}	Deviates
Polyurethane (rigid) foam (PUF)	30–100	2440	2.7×10^{-2}	Deviates
Cellular (Foam concrete)	300–500	983–1570	7 × 10–2	Deviates
Glass fiber	2000	1620	4×10^{-2}	Deviates
Styrofoam	28–34	1340	3.2×10^{-2}	Deviates
Expanded foam (Peripor)	20–30	1240	2.8×10^{-2}	Deviates
Neopor (Graphite foam)	15	1150	3.31×10^{-2}	Deviates

Table 2 A comparative evaluation of various selected insulation materials

Parameter	(EPS)	Fiberglass	Foam concrete	Peripor	Neopor	PUF
Composition	Polystyrene foam	Glass fibro	Cement, foam agent	Polystyrene	Graphite-infused polystyrene	Polyurethane foam
Durability	Long-lasting, with resistance to chemicals and mold	Durable, but may wear over time	Durable	Durable	Durable	Durable
Moisture absorption	Low	Low	Variable	Low	Low	Low
Fire resistance	Variable	Fire-resistant	Variable	Fire-retardant	Fire-retardant	Fire-resistant

1.3 System Simulation and Study Evaluation Framework

Project Overview: Mid-Rise Multifamily Residential Complex

This mid-rise multifamily development encompasses a total shell area of approximately 61,000 square ft and is configured across three above-grade stories. The building is oriented north-facing and features a floor-to-floor height of 9.0 ft, providing 8.0 ft of floor-to-ceiling clearance throughout.

Structural Elements:

Roof assembly:

Constructed with a 6-inch concrete slab, the roof includes light interior insulation but lacks any exterior insulation layer.

Exterior Walls:

The above-grade walls are composed of unfinished concrete and medium asbestos material, with no external cladding or finish applied.

Fenestration:

Windows: Each unit includes standard window openings measuring 4 ft by 4.25 ft.
Doors: Exterior door openings measure 7 ft in height by 6 ft in width.

The building's roof and walls are exposed to external elements like sun and wind, impacting energy performance. Using e-Quest, both 2D and 3D building geometries were modeled to assess energy efficiency. Key parameters such as insulation thickness and material properties were defined, and HVAC systems were configured. Simulations were run across various insulation levels, and results were analyzed to identify the most efficient insulation design for improved energy performance Fig. 2

Life Cycle Cost Analysis (LCCA) is a thorough method for evaluating the most cost-effective solution by analyzing the total costs throughout a system's lifespan. In this analysis, LCCA was used to determine the ideal mix of roofing materials and insulation thickness that optimizes energy efficiency while minimizing future costs. This approach ensures long-term savings and supports sustainable building practices.

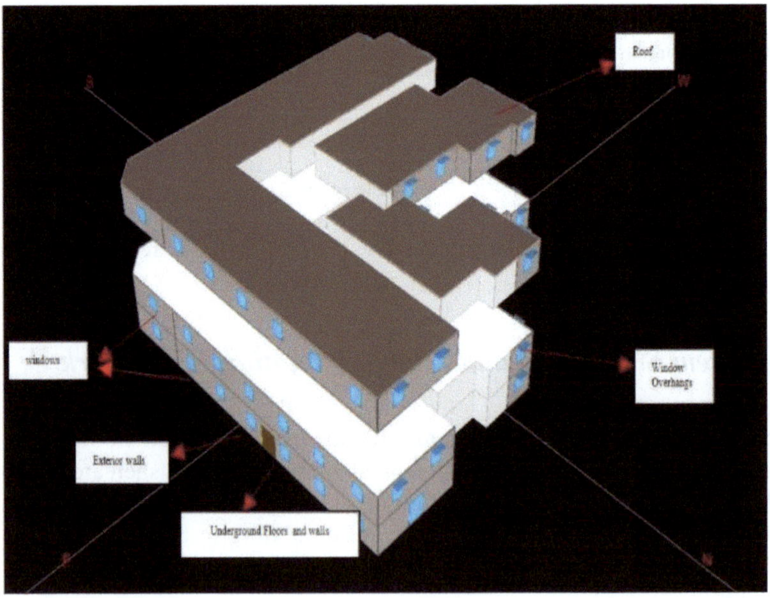

Fig. 2 A 3D model built and simulated using e-Quest software

1.4 Energy Evaluation

The energy analysis examined yearly cooling energy usage and variations in load across different scenarios in various cities in India with diverse climates. The study aimed to identify the most effective combinations of insulation materials and thicknesses to improve energy efficiency and decrease cooling demand. It provided key insights into approaches for optimizing thermal performance while minimizing cooling loads. The total thermal resistance of a multi-layer roof system (Fig. 3) can be calculated using the equation below. These findings contribute to developing practical solutions for energy-efficient building design in varying climatic conditions. The heat conducted through a multi-layer roof as shown in Fig. 3 can be determined through the following equation. Equations (1–3) (Idchabani et al., 2017; Kumar & Suman, 2013; Kallioğlu et al., 2018).

$$\text{Qconducted} = \text{U A } \Delta\text{T}. \tag{1}$$

where in A = roof's surface area

T = thermal differential across the building envelope

U = thermal transmittance (U-value) (W/m^2K).

The thermal resistance provided by a layer of insulation can be determined using the equation below

$$\text{R} = \text{L}/\text{K} \tag{2}$$

And

$$\text{U} = \frac{1}{\sum R_{th}} = \frac{1}{R_1 + R_2 + R_3 + \cdots + R_{n-1} + R_n} \tag{3}$$

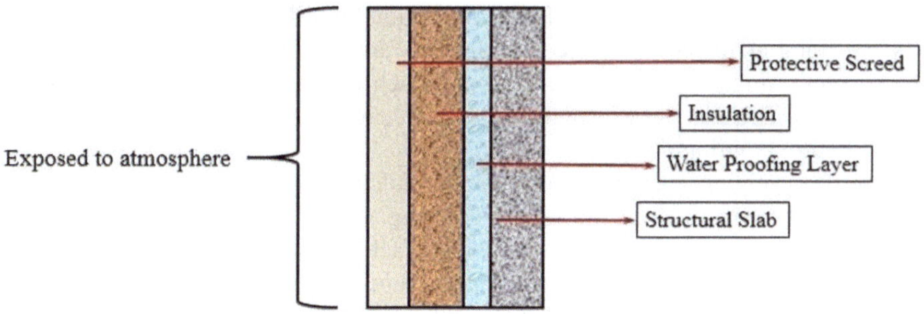

Fig. 3 Cross-sectional representation of roof structure

Table 3 Thermal conductivity resistance values of insulation layers at varying thicknesses across multiple test runs

Thickness of insulation layer (m)	Thermal conductivity resistance values (m^2K/W)					
	Expanded polystyrene	Fiber glass	Foam concrete	Polyurethane foam	Peripor	Neopor
0.01	0.26	0.31	0.20	0.43	0.42	0.36
0.02	0.46	0.44	0.28	0.62	0.60	0.51
0.03	0.72	0.69	0.42	0.99	0.96	0.75
0.04	1.12	1.06	0.63	1.54	1.50	1.27
0.05	1.38	1.31	0.78	1.91	1.86	1.57
0.06	2.04	1.62	1.13	2.84	2.75	2.33
0.07	2.30	2.19	1.28	3.21	3.11	2.63
0.08	2.57	2.44	1.42	3.58	3.47	2.93
0.09	2.70	2.56	1.49	3.77	3.65	3.08
0.1	2.96	2.81	1.64	4.14	4.01	3.38

where R_1, R_2, R_3 ....R_{10} are the thermal conductivity resistance values for Test run 1, Test run 2, Test run 3, and so on, up to Test run 10, are expressed in m^2K/W, with U representing the total thermal conductivity coefficient in W/m^2K and L representing the depth of insulation, measured in meters. Table 3 provides a summary of the thermal resistance for different insulation materials at various thicknesses.

1.5 *Cost Optimization Assessment*

The economic analysis takes into account factors such as interest rates, inflation, and the building's expected lifespan to provide a comprehensive estimate of ongoing energy costs. A customized economic model was developed to assess both the energy expenses and the costs associated with insulation materials. The goal was to identify the optimal scenarios where overall costs are minimized, while still meeting the project's financial objectives. This approach offers valuable insights into the economic consequences of insulation choices, helping to guide the development of cost-efficient building designs.

The life-cycle cost analysis in this study calculates the total cooling expenses over the entire operational life of the building. Electricity rates in various cities across India differ due to several influencing factors, including their distance from energy sources, local infrastructure, government policies, energy mix, supply demand dynamics, utility efficiency, subsidies, and regional economic conditions. These factors contribute to varying pricing structures, as illustrated in Table 8, which lists the rates in Rs/kWh. The Present Worth Factor (PWF) is calculated based on the inflation and interest rates provided in Table 4, reflecting the financial assumptions used in the analysis.

Depending on the lifetime estimated and the present worth factor, the following formulas (Eqs. 4–12) are involved (Annibaldi et al., 2019; Kallioğlu et al., 2023, 2018):

If i > g

$$r = [i-]/[1 + g] \tag{4}$$

If i < g

$$r = [g - i]/[1 + i] \tag{5}$$

$$\mathrm{PVF} = \{[1 + \mathrm{R}] - 1\}/\left\{\mathrm{r}[1 + \mathrm{r}]^{\mathrm{N}}\right\} \tag{6}$$

If i = g

$$\mathrm{PWF} = \mathrm{N}/[I + 1] \tag{7}$$

Total insulation cost is calculated as follows

Table 4 Critical parameters

Description		Values
Life cycle period (N)		10 Years
Resistance of uninsulated wall (R)		0.0617 m^2k/W
Coefficient of performance (COP) cooling		2.5
Total heat transfer area		1043.15 m^2
Rate of interest	i	5.9
Rate of inflation	g	10.7
Present value factor	PVF	4.33
Energy efficiency rate	EER	8.53
Heating effectiveness	η	1
City under study		Cost (Rs/Kw/hr)
Bangalore		7.4
Chennai		8.15
Jaipur		7.98
Kolkata		7.35
Leh		3.35
Mumbai		5.36
New Delhi		7.24
Thiruvananthapuram		7.98

$$C_{ins} = C_{ins} = C_Y \times x \tag{8}$$

$C_{A,C}$ is calculated with the following equations:

$$C_{A,} = [86400 \times CDD \times C_e]/[R_{th}] \times COP] \tag{9}$$

According to the life-cycle cost analysis, the final calculation for the TOTAL cooling cost of an insulated building is as follows.

$$C_{t,C} = C_{A,C} \times PVF + C_Y \times x \tag{10}$$

The ideal insulation thickness, which minimizes overall cooling costs, can be determined using the following equation

Payback period

$$A_{Year,C} = C_C - C_{t,C} \tag{11}$$

Payback Period

$$PP_C = C_{ins} / A_{Year,C} \tag{12}$$

1.6 *Assessment of Sustainability*

The sustainability analysis aims to evaluate the environmental effects of building operations and energy use, focusing particularly on factors such as greenhouse gas (GHG) emissions and temperature reduction. This comprehensive assessment examines how different types of insulation materials and their respective thicknesses influence these critical environmental factors, particularly GHG emissions and the reduction of temperature. To make the findings clearer, a graph has been created to visually display the relationship between insulation thickness, GHG emissions, and temperature reduction for various materials. The results, presented in the findings section, offer insights into patterns and help pinpoint the ideal balance between reducing environmental impact and ensuring effective insulation performance.

2 Results and Discussions

2.1 *Effect of Varying Insulation Thickness on Energy Efficiency*

Roof insulation plays a vital role in enhancing energy efficiency by regulating the internal temperature of a building. In hot climates, it effectively prevents external heat from penetrating, keeping indoor spaces cooler, while in colder climates, it helps retain the warmth inside, reducing heating requirements. Among the insulation materials evaluated, polyurethane foam (PUF) emerges as the most effective in reducing electricity consumption. Other materials like Peripor, Neopor, EPS, and Fiberglass also contribute to energy savings, although their impact is not as significant as PUF. On the other hand, Foam Concrete proves to be the least efficient in terms of energy savings, offering limited benefits when it comes to reducing overall energy usage.

Cities like Bangalore and Chennai benefit the most from roof insulation, experiencing the highest reductions in energy use. Bangalore, in particular, shows the greatest savings across all insulation materials, indicating its strong effectiveness in this region. Other cities, including New Delhi, Bangalore, Kolkata, Mumbai, Jaipur, Thiruvananthapuram, and Leh, demonstrate a diverse range of energy savings, with certain cities showing more modest reductions in energy consumption. The varying effectiveness across regions, such as Bangalore and Chennai, which show the highest savings, highlights how local climate conditions influence the performance of insulation. This demonstrates that the benefits of insulation can differ significantly based on the specific temperature patterns and environmental factors of each region (Table 5).

2.2 *Effect of Insulation Thickness on the Ct,c and CA,c*

Increasing insulation thickness typically reduces the annual cooling cost by improving the building's thermal resistance. Thicker insulation slows down heat transfer, keeping indoor temperatures more stable and reducing the need for air conditioning. This leads to lower energy consumption for cooling, thereby decreasing cooling costs. However, the reduction in cooling costs may diminish after a certain thickness, as the incremental benefits of added insulation decrease with greater thickness. Increasing the thickness of insulation enhances thermal resistance, resulting in a reduction of total cooling costs and lower energy consumption for cooling. However, this must be balanced with the additional cost of the insulation material itself. While thicker insulation lowers energy usage and cooling expenses, the initial cost of more material can offset some of the savings. The net effect on total cooling costs depends on the balance between reduced energy consumption and the added upfront cost of the thicker insulation. At a certain point, the added material cost may outweigh the savings in cooling energy.

Table 5 Range of electricity savings achieved through different insulation materials in trials conducted across eight Indian cities

Reduction in electricity use (percentage range)								
Insulation material	Indian cities							
	Bangalore	Chennai	Jaipur	Kolkata	Leh	Mumbai	New Delhi	Thiruvananthapuram
EPS	6.51–11.63	5.52–9.63	5.12–8.82	5.3–8.68	1.38–1.71	5.68–10.01	5.76–10.29	5.0–8.24
Fiber glass	6.33–11.59	5.39–9.6	4.99–8.79	5.18–8.65	1.35–1.71	5.53–9.97	5.6–10.25	4.87–8.22
Foam concrete	4.52–10.59	3.97–8.95	3.59–8.4	3.87–8.34	1.07–1.70	3.97–9.5	4.03–9.71	3.63–7.88
PUF	7.64–11.49	6.38–9.81	6.06–8.96	6.08–8.79	1.5–1.72	6.79–10.18	6.8–10.51	5.79–8.37
Peripor	7.52–11.84	6.31–9.75	5.99–8.96	6.01–8.78	1.49–1.72	6.71–10.17	6.72–10.49	5.72–8.35
Neopor	6.96–11.73	5.87–9.71	5.46–8.89	5.62–8.73	1.4–1.71	6.24–10.08	6.15–10.39	5.32–8.30

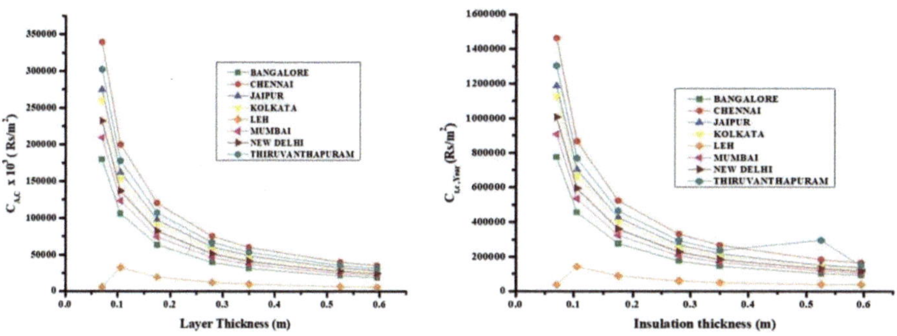

(a): $C_{A,C}$ versus insulation thickness

(b) : $C_{t,C}$ versus insulation thickness

Fig. 4 Correlation between insulation thickness and yearly cooling cost

Figure 4 clearly illustrates that increasing insulation thickness leads to a reduction in both annual energy expenses and overall cooling costs. However, beyond a certain point, the graph levels off, indicating that further increases in insulation thickness no longer lead to significant reductions in cooling costs, but only contribute to higher overall costs due to the added material. Figures (a) and (b) illustrate this trend across all cities, with the exception of Leh for annual cooling costs and Jaipur for total cooling costs. This deviation can likely be attributed to the unique climatic conditions in these regions, which influence the effectiveness of insulation at different thicknesses.

2.3 Insulation Thickness and Its Role in Reducing GHG Emissions and Improving Temperature Control

The type of insulation used in a building greatly impacts its fuel consumption. Selecting appropriate insulation materials helps minimize thermal loss through the building's envelope. Moreover, using the right insulation reduces emissions, as fuel energy is converted into harmful gases like CO_2, SO_2, CO, and others. The decrease in CO_2 emissions for coal as a fuel type is illustrated in the figure. Figure 5 shows a notable reduction in CO_2 emissions for the cities under study, ranging from 95% to 97% when considering sustainable and green building practices.

As the thickness of insulation increased, greenhouse gas emissions decreased, with reductions ranging from 80% to 90%. However, Fig. 6 indicates that this decline eventually plateaued. When coal is used as the fuel for electricity generation, CO_2 emissions across all selected Indian cities range between 6.92 and 10.67 kg/m^2 and 0.014–0.002 Kg/m^2, respectively.

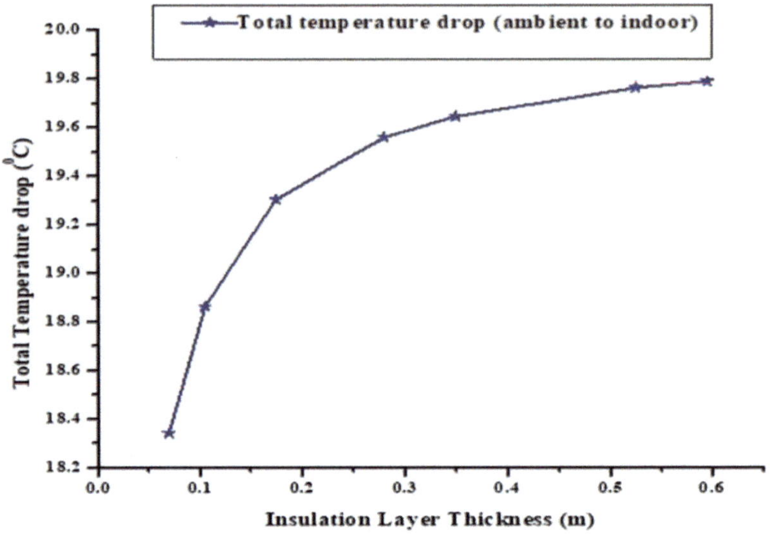

Fig. 5 Temperature drop vs insulation layer thickness

Fig. 6 CO_2 emissions for coal as fuel

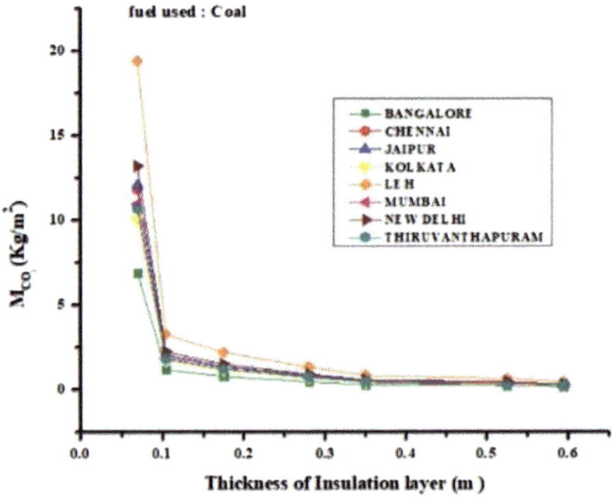

3 Conclusion

3.1 Key Insights

- Polyurethane (Rigid) Foam (PUF) proves to be the most efficient insulation material among those analyzed, offering substantial improvements in energy efficiency and reducing electricity consumption by approximately 33% to 97%, depending

on the specific conditions and application. In contrast, Foam Concrete exhibits the least effective performance, with energy savings ranging from 31 to 93%. This comparison highlights PUF's superior thermal insulation capabilities and its greater potential to lower energy usage compared to other materials tested.
- Bangalore stands out as the most responsive city, showing significant and consistent reductions in energy consumption due to insulation. In contrast, Leh displays fluctuating results, with energy savings largely dependent on its unique regional climate. This indicates that insulation effectiveness can vary greatly with local environmental conditions, requiring tailored approaches for optimal performance.
- As the insulation thickness increased, greenhouse gas emissions decreased, with a reduction ranging from 80 to 90%.

3.2 Key Focus Areas for Ongoing Research and Development

- Integration with Building Systems: Future studies can investigate how roof insulation can be combined with other building systems—such as HVAC, lighting, and renewable energy technologies—to enhance overall energy efficiency and building performance. This integrated approach could lead to smarter energy management and greater long-term sustainability in building operations.
- Advancements in Energy Modeling: Improving energy modeling methods will enable more accurate forecasts of thermal behavior across diverse climatic conditions, leading to more informed design choices and increased building energy efficiency. Enhanced modeling can also help evaluate the long-term impact of insulation strategies, ensuring buildings are both cost-effective and environmentally sustainable.
- Education and Awareness: Educating architects, engineers, contractors, and building owners on the proper selection, installation, and upkeep of roof insulation systems can significantly improve building efficiency and reduce energy usage. Increased awareness ensures that insulation is applied effectively, maximizing its thermal benefits and contributing to long-term operational savings and sustainability goals.

References

Abdelzaher, M. A. (2023). Sustainable development goals for industry, innovation, and infrastructure: demolition waste incorporated with nanoplastic waste enhanced the physicomechanical properties of white cement paste composites. *Applied Nanoscience (Switzerland), 13*(8), 5521–5536. https://doi.org/10.1007/s13204-023-02766-w

Aktemur, C., Bilgin, F., & Tunçkol, S. (2021). Optimisation on the thermal insulation layer thickness in buildings with environmental analysis: an updated comprehensive study for Turkey's all provinces. *Journal of Thermal Engineering, 7*(5), 1239–1256. https://doi.org/10.18186/thermal.978057

Anand, Y., Anand, S., Gupta, A., & Tyagi, S. K. (2015). Building envelope performance with different insulating materials - An exergy approach. *Journal of Thermal Engineering, 1*(4), 433–439. https://doi.org/10.18186/jte.06871

Annibaldi, V., Cucchiella, F., Berardinis, P. De, Rotilio, M., & Stornelli, V. (2019). Environmental and economic benefits of optimal insulation thickness : A life-cycle cost analysis. Renewable and Sustainable Energy Reviews, 116(September), 109441. https://doi.org/10.1016/j.rser.2019.109441

Boixo, S., Diaz-Vicente, M., Colmenar, A., & Castro, M. A. (2012). Potential energy savings from cool roofs in Spain and Andalusia. *Energy, 38*(1), 425–438. https://doi.org/10.1016/j.energy.2011.11.009

Dong, Y., Kong, J., Mousavi, S., Rismanchi, B., & Yap, P.-S. (2023). Wall Insulation Materials in Different Climate Zones: A Review on Challenges and Opportunities of Available Alternatives. *Thermo, 3*(1), 38–65. https://doi.org/10.3390/thermo3010003

Friess, W. A., & Rakhshan, K. (2017). A review of passive envelope measures for improved building energy efficiency in the UAE. *Renewable and Sustainable Energy Reviews, 72*(November 2016), 485–496. https://doi.org/10.1016/j.rser.2017.01.026

González-Torres, M., Pérez-Lombard, L., Coronel, J. F., Maestre, I. R., & Yan, D. (2022). A review on buildings energy information: Trends, end-uses, fuels and drivers. *Energy Reports, 8*(December 2021), 626–637. https://doi.org/10.1016/j.egyr.2021.11.280

Halwatura, R. U., & Jayasinghe, M. T. R. (2008). *Thermal performance of insulated roof slabs in tropical climates. 40*, 1153–1160. https://doi.org/10.1016/j.enbuild.2007.10.006

Idchabani, R., Khyad, A., & El Ganaoui, M. (2017). Optimizing insulation thickness of external walls in cold region of Morocco based on life cycle cost analysis. *Energy Procedia, 139*, 117–121. https://doi.org/10.1016/j.egypro.2017.11.183

Kallioğlu, M. A., Ercan, U., Avcı, A. S., Fidan, C., & Karakaya, H. (2018). *The Economical and Environmental Analysis of Optimum Insulation Thickness Within the Process of Heating and Cooling. October*. https://doi.org/10.20944/preprints201810.0070.v1

Kallioğlu, M. A., Yılmaz, A., Sharma, A., Mohamed, A., Dobrotă, D., Alam, T., Khargotra, R., & Singh, T. (2023). Optimal Insulation Assessment, Emission Analysis, and Correlation Formulation for Indian Region. *Buildings, 13*(2). https://doi.org/10.3390/buildings13020569

Karpov, D., Dyudina, O., & Pavlov, M. (2021). A review on modern heat-insulating materials for improving the energy efficiency of buildings and life-support utilities. *E3S Web of Conferences, 288*, 01099. https://doi.org/10.1051/e3sconf/202128801099

Kaynakli, O. (2008). A study on residential heating energy requirement and optimum insulation thickness. *Renewable Energy, 33*(6), 1164–1172. https://doi.org/10.1016/j.renene.2007.07.001

Kumar, A., & Suman, B. M. (2013). Experimental evaluation of insulation materials for walls and roofs and their impact on indoor thermal comfort under composite climate. *Building and Environment, 59*, 635–643. https://doi.org/10.1016/j.buildenv.2012.09.023

Mahapatra, D. (2022). *Evaluation of Optimum Thickness of Insulation Materials and Their Carbon Mitigation Potential in Indian Climatic Zone.*

Mahlia, T. M. I., Taufiq, B. N., Ismail, & Masjuki, H. H. (2007). Correlation between thermal conductivity and the thickness of selected insulation materials for building wall. *Energy and Buildings, 39*(2), 182–187. https://doi.org/10.1016/j.enbuild.2006.06.002

MD Jalal, A Tanveer, K. J. and F. A. (1991). Foam concrete. *Concrete (London), 25*(1), 49–54. http://www.ripublication.com

Mirrahimi, S., Mohamed, M. F., Haw, L. C., Ibrahim, N. L. N., Yusoff, W. F. M., & Aflaki, A. (2016). The effect of building envelope on the thermal comfort and energy saving for high-rise buildings in hot-humid climate. *Renewable and Sustainable Energy Reviews, 53*, 1508–1519. https://doi.org/10.1016/j.rser.2015.09.055

Mishra, S., Usmani, J. A., & Varshney, S. (2012). Energy Saving Analysis In Building Walls Through Thermal Insulation System. *Sanjeev Varshney / International Journal of Engineering Research and Applications (IJERA), 2*(5), 128–135.

Pearlmutter, D., & Rosenfeld, S. (2008). Performance analysis of a simple roof cooling system with irrigated soil and two shading alternatives. *Energy and Buildings*, *40*(5), 855–864. https://doi.org/10.1016/j.enbuild.2007.06.004

Rajak, D. K., Wagh, P. H., & Linul, E. (2021). Manufacturing technologies of carbon/glass fiber-reinforced polymer composites and their properties: A review. *Polymers*, *13*(21). https://doi.org/10.3390/polym13213721

Ran, J., Tang, M., Jiang, L., & Zheng, X. (2017). Effect of Building Roof Insulation Measures on Indoor Cooling and Energy Saving in Rural Areas in Chongqing. *Procedia Engineering*, *180*, 669–675. https://doi.org/10.1016/j.proeng.2017.04.226

Raut, A., Khatoon, S., & Goud, P. (2019). A comparative study on effects of various insulating layers of roof system on energy usage of building envelope. *IOP Conference Series: Earth and Environmental Science*, *354*(1). https://doi.org/10.1088/1755-1315/354/1/012055

Shrimali, R., & Agrawal, N. K. (2024). Climate responsive insulation strategies: a comparative analysis for enhanced energy conservation and reduced environmental footprint in Indian urban contexts. In *Environment, Development and Sustainability* (Issue 0123456789). Springer Netherlands. https://doi.org/10.1007/s10668-024-05267-0

Shrimali, R., Agrawal, N. K., & Agrawal, K. K. (2024). Energy and cost analysis of composite-insulated rooftops for sustainable buildings in Indian climatic conditions. *Energy Sources, Part B: Economics, Planning and Policy*, *19*(1). https://doi.org/10.1080/15567249.2024.2316729

Simona, P. L., Spiru, P., & Ion, I. V. (2017). Increasing the energy efficiency of buildings by thermal insulation. *Energy Procedia*, *128*, 393–399. https://doi.org/10.1016/j.egypro.2017.09.044

Singh, H. K., Prakash, R., & Shukla, K. K. (2016). *{Economic} {Insulation} {Thickness} {of} {External} {Walls} {in} {Hot} {and} {Composite} {Regions} {of} {India}*. *March 2020*, 6.

Wang, H., Huang, Y., Guo, C., Yang, L., & Huang, L. (2023). A low-carbon optimization design method for building roof insulation using comprehensive economic analysis model and evaluation index. *Sustainable Energy Technologies and Assessments*, *55*(August 2022), 102937. https://doi.org/10.1016/j.seta.2022.102937

Efficient Image Captioning with XceptionNet and Nyströmformer

Anjali Sharma and Mayank Aggarwal

Abstract Image description generation is an intricate process incorporating computer vision and natural language processing to produce descriptive captions for photographs. This paper presents a novel architecture combining XceptionNet as the image encoder with Nyströmformer, a linear-complexity transformer variant, as the caption decoder. By leveraging linear attention mechanisms, the proposed model mitigates the quadratic complexity of traditional transformers, ensuring efficient and scalable caption creation. Evaluated on the Microsoft Common Objects in Context dataset using BLEU, ROUGE, and CIDEr, the model demonstrates competitive performance while significantly reducing computational overhead. Additionally, a mathematical discussion of the Nyströmformer's attention mechanism highlights its advantages over conventional transformer-based approaches.

Keywords Transformers · XceptionNet · Nyströmformer · Linear attention mechanisms · Image captioning

1 Introduction

Image description generation is a fundamental task in artificial intelligence that bridges computer vision and natural language processing by producing relevant textual descriptions from visual content (Al-Shamayleh et al. 2024). It has numerous real-world applications, including assistive technologies, autonomous systems, and content retrieval. Recent developments in deep learning have led to the development of powerful architectures that combine transformers for caption creation and convolutional neural networks (CNNs) for deriving features. However, conventional transformer-based models suffer from quadratic complexity in regard to sequence

A. Sharma (✉) · M. Aggarwal
Computer Science and Engineering, FET, Gurukula Kangri (Deemed to be) University, Uttrakhand, India
e-mail: 23631001@gkv.ac.in

M. Aggarwal
e-mail: mayank@gkv.ac.in

M. Ram et al. (eds.), *Technological Advancements in Engineering and Management Sciences*, Management and Industrial Engineering,
https://doi.org/10.1007/978-3-032-04606-2_11

length, making them computationally expensive for long captions and large-scale datasets (Tang 2024).

To tackle these issues, this work suggests a new image captioning framework that integrates XceptionNet as the image encoder and Nyströmformer as the caption decoder. XceptionNet (Khodra 2024), known for its separable depthwise convolutions, efficiently extracts rich image features. At the same time, Nyströmformer, a variant of a linear-complexity transformer, mitigates the high computational cost of self-attention mechanisms. Unlike traditional transformers, Nyströmformer reduces memory overhead while maintaining competitive proficiency, making it suitable for scalable caption generation (Xiong 2021; Cui et al. 2018).

The suggested model uses the MS COCO dataset using standard NLP metrics such as BLEU, ROUGE, and CIDEr, demonstrating comparable caption effectiveness with significantly reduced computational costs. This work highlights the potential of linear attention mechanisms in image captioning and provides a mathematical discussion of the efficiency of Nyströmformer compared to other linear transformers. The findings suggest that integrating computationally efficient transformers may make scalable and high-quality caption creation possible, opening the door to developments in multimodal AI applications.

2 Related Work

2.1 CNNs for Image Captioning

Image captioning requires a smooth amalgamation of computer vision and natural language processing to provide logical and contextually appropriate textual captions of photos. This topic has significantly advanced due to deep learning architectures, especially Convolutional Neural Networks (CNNs) and Recurrent Neural Networks (RNNs) (Hoxha et al. 2020). Because of its effectiveness, depthwise separable convolutions, and ability to capture fine features in images, XceptionNet has become one of the most potent feature extractors among them.

- **XceptionNet for Image Captioning**: A popular version of the Inception architecture for feature extraction in image captioning tasks is called XceptionNet. It collects high-level semantic properties from images, analyzes them efficiently, and uses those qualities to provide meaningful written descriptions. The model performs better on common assessment criteria such as BLEU and METEOR and increases the quality of the caption due to its ability to capture complex visual information (Pardede and Fandi 2024). When used with sequence-based models, Xception's deep feature extraction capabilities perform very well. These models use the retrieved characteristics as input, allowing a precise and logical caption creation. Research has shown that merging Xception with LSTM greatly enhances caption creation using the long-term dependencies of sequential data (Kanimozhiselvi et al. 2022).

- **Integration of Xception with LSTM and Transformer Models**: Combining Xception with LSTM networks is a popular method for image captioning. This method creates captions by processing high-dimensional image embeddings that Xception extracted using an LSTM decoder. This combination improves semantic coherence and context preservation, resulting in more descriptive and expressive captions (Joshi et al. 2024). More recently, Xception has been combined with Transformer-based models to generate captions. Transformers with self-attention processes enhance long-range dependencies and context knowledge in text creation. Transformer decoders can increase caption accuracy by dynamically focusing on pertinent portions of the image's characteristics using attention. This method has proven to perform better than others when dealing with complicated photographs that require precise descriptions of numerous items and minute features (Kurlekar 2023).

Integrating Xception with advanced attention models, such as Nyströmformer, presents a promising direction for achieving more efficient and scalable caption generation. Further studies are required to explore hybrid architectures that leverage linear transformers for improved efficiency and generalization across different datasets.

2.2 *Transformers in Image Captioning*

Transformers have become central to modern image captioning systems because they can model long-range dependencies via self-attention. Traditional approaches relied on encoder-decoder frameworks with CNNs and RNNs, but transformer-based models now dominate, offering superior performance through multi-head attention and parallel processing (Ondeng et al. 2023). Recent advancements, such as vision-language pre-training (VLP), enhance caption quality by leveraging large multimodal datasets. Vision transformer models treat images as sequences of patches, allowing for better global feature extraction compared to traditional convolutional methods (Navalkar and Wankhade 2024). Additionally the use of attention in transformers enables the model to pay attention on prominent image sections, enhancing the meaning of generated captions (Nabi et al. 2023). Models like ETransCap prioritize efficiency for applications requiring real-time caption generation (e.g., robotics or accessibility tools), achieving CIDEr scores comparable to cutting-edge models while using fewer resources (Mundu et al. 2024).

Traditional transformers struggle with lengthy sequences due to their quadratic scaling in self-attention $O(n^2)$. In image captioning, this becomes critical when processing high-resolution images split into hundreds of patches (e.g., via Vision Transformers (ViT)) or handling dense captions. Linear transformers approximate attention mechanisms using low-rank projections or landmark token selection, reducing complexity to O(n) (Fan and Li 2023). Different linear transformer models, such as

Table 1 Research gaps in efficient transformers and Nyströmformer's advancements

Model	Research gap	Motivation for Nyströmformer
Reformer (Yang et al. 2022)	LSH-based hashing fails for non-local dependencies; chunking disrupts context	Landmark tokens preserve global context without chunking
Performer (Lan 2023)	Kernel approximations introduce variance, degrading softmax accuracy	Matrix decomposition retains softmax fidelity
Longformer (Sun et al. 2024)	Sparse attention limits global interaction; fixed patterns reduce adaptability	Landmarks sample full sequence context dynamically
Linformer (Zhang et al. 2021)	Fixed low-rank projections lose fine-grained token interactions	Adaptive landmark selection avoids rigid rank reduction

Linformer, Reformer, and Performer, are prevalent in research for image captioning and have proven efficient.

Table 1 highlights the limitations of prior work and motivations for adopting Nyström-based attention in image captioning.

Along with the above advantages, Nyströmformer combines linear complexity (O(n)) with softmax-compatible global attention, which has proven to be a solution to existing image captioning systems.

Figures 1 and 2 visualization compares the Computational Complexity (Time and memory) and Convergence Speed of different linear transformers (Yang et al. 2022; Lan 2023; Sun et al. 2024; Zhang et al. 2021). Nyströmformer offers a balanced trade-off between efficiency and accuracy, while Performer is the fastest. Longformer is the most computationally expensive due to its local-global attention mechanism.

3 Proposed Architecture

The proposed framework consists of two primary modules:

- **Image Encoder**: XceptionNet with linear attention to efficiently extract and process image features.
- **Caption Decoder**: Nyströmformer, an attention approximation-based transformer, to generate high-quality textual descriptions.

The block diagram for the suggested image captioning system architecture is displayed in Fig. 3.

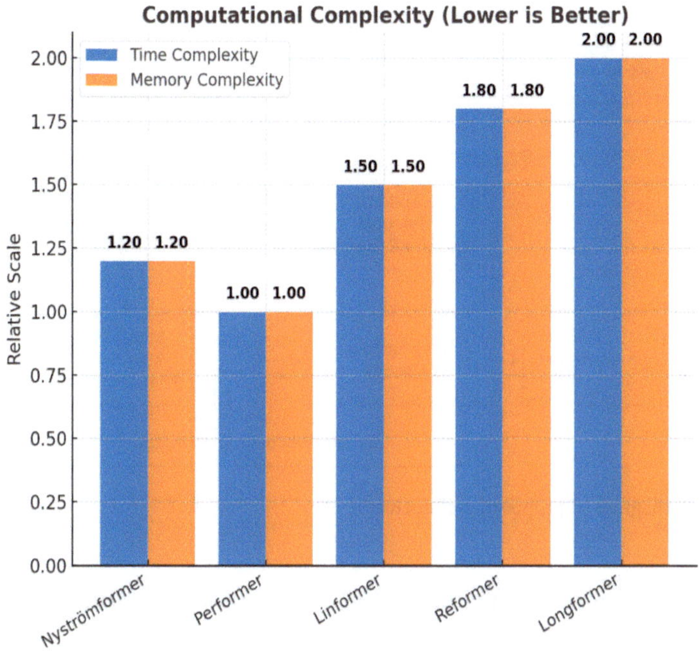

Fig. 1 Comparative analysis of linear transformers for computational complexity

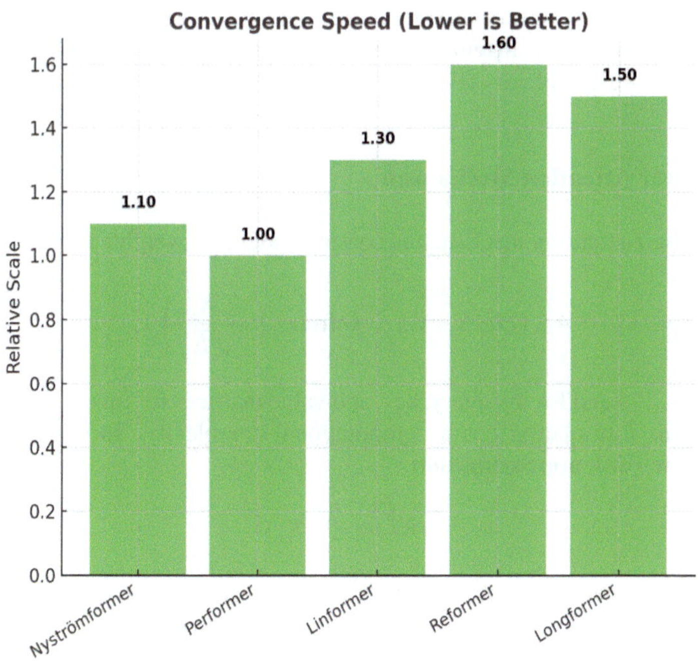

Fig. 2 Comparative analysis of linear transformers for convergence speed

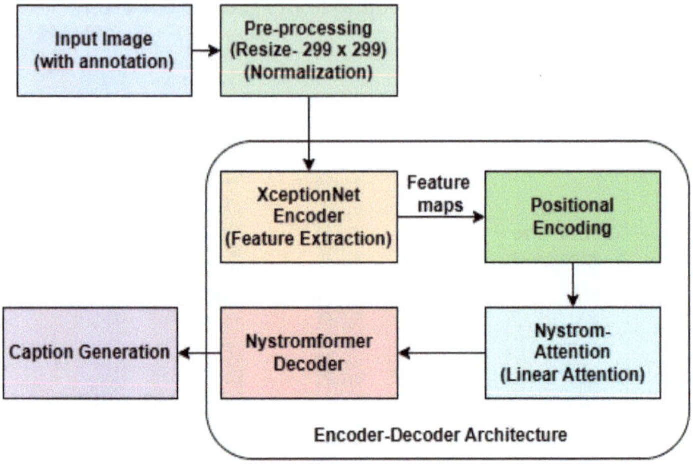

Fig. 3 Block diagram of proposed architecture

3.1 Image Encoder: XceptionNet with Linear Attention

XceptionNet is a deep convolutional neural network that replaces standard convolutions with **depthwise separable convolutions**, reducing computational overhead while maintaining high performance. The extracted feature representations are further refined using a **linear attention mechanism**, reducing standard attention mechanisms' quadratic complexity (Fudholi et al. 2022).

3.1.1 Linear Attention Mechanism

Traditional self-attention mechanisms compute attention weights using:

$$A_{ttn}(Q_i, K_i, V_i) = \text{softmax}\left(\frac{Q_i K_i^T}{\sqrt{d_k}}\right) V_i \tag{1}$$

where Q, K, V stand for the query, key, and value matrices for image i, respectively. However, this formulation has a computational complexity. To mitigate this, we employ a **low-rank approximation**:

$$Q_i K_i^T \approx Q_i'(K_i')^T \tag{2}$$

where and are low-rank approximations of and, respectively. This approximation reduces the complexity, allowing scalable attention computation.

3.2 *Caption Decoder: Nyströmformer*

The **Nyströmformer** is an efficient transformer variant that approximates the self-attention mechanism using the **Nyström method**, reducing quadratic complexity. The decoder takes image feature representations from the encoder and generates descriptive captions (Ning 2024).

3.2.1 Nyströmformer Attention Mechanism

Given an input sequence, self-attention is computed as:

$$A = \text{softmax}\left(\frac{QK^T}{\sqrt{d_k}}\right) \tag{3}$$

To approximate using **Nyström approximation**, a subset of landmark points is chosen, and the approximation is given by:

$$A \approx \tilde{A} = \tilde{Q}\tilde{K}^T(\tilde{K}\tilde{K}^T)^{-1}\tilde{K}Q^T \tag{4}$$

where Q and K are query and key matrices for landmark points. This approximation reduces computational complexity, enabling efficient processing of long sequences.

3.3 *Algorithm for the Proposed Framework*

The given algorithm shows the complete suggested model for Image Captioning with XceptionNet and Nyströmformer.

Image Captioning with XceptionNet and Nyströmformer
Given an input image I and a maximum caption length T, the algorithm generates a caption sequence $C = \{C_1, C_2, \ldots, C_T\}$ using the following steps:

- Extract image features using XceptionNet: $F \Leftarrow \text{XceptionNet}(I)$
- Compute queries, keys, and values: $Q, K, V \Leftarrow W_Q F, W_K F, W_V F$
- Approximate QK^T using low-rank approximation: $Q'K'^T \approx QK^T$
- Compute linear attention: $A' \Leftarrow \text{softmax}\left(\frac{Q'K'^T}{\sqrt{d_k}}\right) V$
- Initialize caption sequence with $C_0 \Leftarrow$ <start>

Then, for each time step $t = 1$ to T:

- Compute new Q_t, K_t, V_t from the previous token C_{t-1}
- Use Nyström approximation for attention:

$$A_t \approx \tilde{A}_t = \tilde{Q}_t \tilde{K}_t^T (\tilde{K}_t \tilde{K}_t^T)^{-1} \tilde{K}_t Q_t^T$$

- Compute output: $O_t \Leftarrow A_t V_t$
- Predict next token: $C_t \Leftarrow \arg\max \text{softmax}(W_o O_t)$
- If C_t is <end>, stop

Finally, return the caption sequence $C = \{C_1, C_2, \ldots, C_t\}$.

4 Methodology

4.1 Dataset

Table 2 summarizes the division of the MS COCO dataset used to train and evaluate the suggested model, with a total of 120,062 images divided between training, validation, and test sets.

Table 2 MS COCO dataset split for training and evaluation

Dataset split	Number of photographs
Image set for training	82,783
Image set for validation	40,504
Image set for testing	40,775
Total images	120,062

4.2 Preprocessing

The images are first scaled to 299 $\times$ 299 pixels to match the input size of XceptionNet. They are then normalized using the mean and standard deviation of the ImageNet dataset. The captions undergo tokenization, then are converted to lowercase, and are augmented with unique tokens to mark the start (`_START_`) and end (`_END_`) of each caption.

4.3 Training

The model was constrained using cross-entropy loss and the Adam optimizer. In training, a learning rate of 1e-4 and a batch size of 8 are used to ensure better results. There are two steps in the training process: the first uses cross-entropy loss to maximize the CIDEr score. Table 3 displays the hyperparameter settings used to train the proposed model.

4.4 Implementation Details

The suggested model is implemented using PyTorch, leveraging pre-trained architectures and efficient libraries for deep learning. The image encoder is based on **XceptionNet**, loaded from the `torchvision` library. At the same time, the caption decoder is implemented using the **Nyströmformer** architecture, available in the `Hugging Face Transformers` library. Model training is performed on a single NVIDIA V100 GPU with 32GB of memory, ensuring efficient computation and faster convergence.

Table 3 Hyperparameter settings for model training

Parameter	Value
Learning rate	1×10^{-4}
Size of batch	8
Number of epochs	30
Dropout rate	0.1
Nyströmformer layer count	6
Attention head count	8
Size of the hidden layer	512
Beam size	5

4.4.1 Image Encoder Implementation

The image encoder is implemented using the **XceptionNet** architecture. To tailor the model for image captioning, the final fully connected layer is replaced with a linear layer to reduce the dimensionality of the extracted feature vector. The processed feature vector is then fed into the Nyströmformer decoder for caption generation.

4.4.2 Caption Decoder Implementation

The caption decoder follows the **Nyströmformer** architecture, which approximates the mechanism of self-attention through the Nyström method. This approach enables efficient handling of long sequences while maintaining high-quality text generation. The decoder monitors the image features extracted by the encoder, progressively generating descriptive captions.

4.4.3 Training Loop

The model training follows a structured sequence of operations to optimize caption generation:

1. **Forward Pass**: A feature vector is derived from the input picture and processed via the XceptionNet encoder. The matching caption tokens are passed through the Nyströmformer decoder to produce an output sequence.
2. **Loss Calculation**: The cross-entropy loss function compares the produced captions with the reference captions to ensure accurate predictions.
3. **Backward Pass**: Gradients are calculated by backpropagation to improve learning efficiency, and the Adam optimizer is used to modify the model parameters.

This structured approach ensures stable and practical model training, leading to improved proficiency in image captioning jobs.

4.5 Evaluation Metrics

The competence of the suggested model is tested using widely accepted metrics for image captioning, ensuring a comprehensive assessment of generated captions. The following metrics are calculated:

- **BLEU (Bilingual Evaluation Understudy)**: The Bilingual Evaluation Understudy, or BLEU, compares the produced captions against reference captions to determine their accuracy and assesses the produced and ground-truth captions' n-gram overlap (Papineni et al. 2002).

- **ROUGE (Recall-Oriented Understudy for Gisting Evaluation)**: Calculate the recall of the captions produced by comparing them to the reference captions. It is beneficial for summarization and text generation tasks (Graham 2015).
- **CIDEr (Consensus-based Image Description Evaluation)**: Measures the consensus among the produced and original captions by taking into account the frequency of n-grams, emphasizing words more relevant to human descriptions (Young et al. 2023).

These evaluation metrics ensure a robust and fair assessment of the model's proficiency in producing human-like descriptions from images.

5 Experimental Results

The implementation of the model was carried out using Python and required a GPU for efficient training and evaluation. The model attained a final training loss of 1.2537 with 84.84% accuracy, confirming effective model learning. The validation loss of 1.1525 and an accuracy of 84.57% suggest strong generalization capabilities. Additionally, the final test accuracy was 83. 31% demonstrates the strength of the model on unseen data. The model size of 225.71 MB reflects its complexity and computational capacity. Figure 4 presents two plots: Loss per Epoch and Accuracy per Epoch during training. The first plot exhibits a decreasing trend in training and validation loss, while the second plot highlights stable accuracy after an initial improvement phase.

Figure 5 illustrates two line charts depicting the convergence of validation loss and sensitivity of validation accuracy over training epochs. The left graph demonstrates a rapid decline in convergence, signifying stabilization in loss, while the right chart

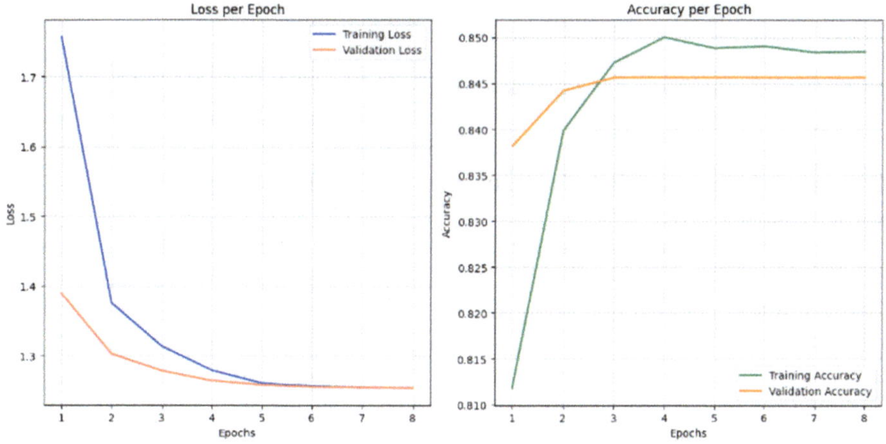

Fig. 4 Accuracy and loss during epochs

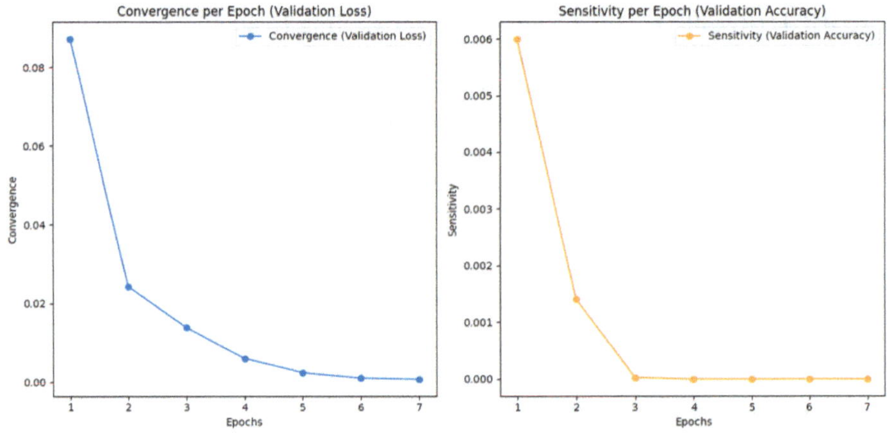

Fig. 5 Convergence and sensitivity plots

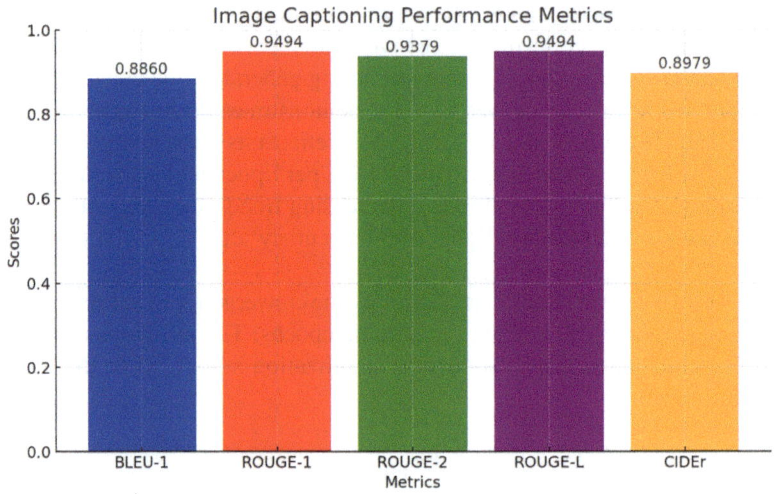

Fig. 6 The evaluation metrics of the proposed model

shows a decreasing sensitivity, indicating that validation accuracy remains consistent across epochs.

The suggested model produced comparable results using the following metrics: BLEU-1: 0.886, ROUGE-1: 0.9494, ROUGE-2: 0.9379, ROUGE-L: 0.9494, and CIDEr: 0.8979, as illustrated in Fig. 6.

Table 4 shows sample picture captions produced using the suggested model compared to reference captions. Every entry consists of a picture, captions created by the suggested model, and a human-annotated original caption. The findings demonstrate how well the model can replicate human descriptions, although with little word choice and phrasing changes.

Table 4 Sample image captions generated by the model

Image with captions

Reference caption: A black dog on leash and people's legs Generated caption: A black dog on a leash and people's legs them

Reference caption: A city bus carrying featuring women coming to town Generated caption: A city bus featuring women coming to town

6 Conclusion

This research introduces a new image captioning framework that uses Nyströmformer, a computationally fast decoder with linear attention mechanisms, and XceptionNet for feature extraction. With an accuracy of 84.84% for training, 84.57% for validation, and 83.31% for test data, the proposed model shows robustness and strong generalization and performs competitively on the MS COCO dataset. With a model size of 225.71 MB, the architecture successfully strikes a compromise between accuracy and scalability, making it appropriate for real-world deployment. Convergence and sensitivity analysis complement the experimental results, demonstrating a steady accuracy and a continuous drop in loss throughout training epochs. The model's ability to produce high-quality captions is further supported by its achievement of state-of-the-art captioning measures, including BLEU-1 (0.886), ROUGE-1 (0.9494), and CIDEr (0.8979).

References

Al-Shamayleh AS, Adwan O, Alsharaiah MA, Hussein AH, Kharma QM, Eke CI (2024) A comprehensive literature review on image captioning methods and metrics based on deep learning technique. Multimedia Tools Appl 83(12):34219–34268. https://doi.org/10.1007/s11042-024-18307-8

Cui Y, Yang G, Veit A, Huang X, Belongie S (2018) Learning to evaluate image captioning. In: Proc IEEE Conf Comput Vis Pattern Recognit, pp 5804–5812

Fan X, Li Y (2023) An Improved Informer Network for Short-Term Electric Load Forecasting. Recent Adv Electr Electron Eng 16(5):532–540

Fudholi DH, Zahra A, Nayoan RAN (2022) A study on visual understanding image captioning using different word embeddings and CNN-based feature extractions. Kinetik: Game Technol Inf Syst Comput Netw Comput Electron Control:91–98
Graham Y (2015) Re-evaluating automatic summarization with BLEU and 192 shades of ROUGE. In: Proc Conf Empir Methods Nat Lang Process, pp 128–137
Hoxha G, Melgani F, Slaghenauffi J (2020) A new CNN-RNN framework for remote sensing image captioning. In: M2GARSS, pp 1–4
Joshi A, Kalal K, Bhandare D, Patil V, Kulkarni U, Meena SM (2024) Image captioning using CNN-LSTM. In: Lect Notes Electr Eng. Springer, Singapore, pp 421–433
Kanimozhiselvi CS, Karthika KS, Krithika (2022) Image captioning using deep learning. In: 2022 Int Conf Comput Commun Informatics (ICCCI), Coimbatore, India
Khodra M (2024) Image Captioning With Sentiment for Indonesian. In: 2024 Int Conf Artif Intell Blockchain Cloud Comput Data Anal (ICoABCD), IEEE
Kurlekar S (2023) Feature extraction using CNN for image captioning. In: 40th GISFI Stand Ser Meet Joint Int Conf 6G Wirel Netw Technol
Lan E (2023) Performer: A novel PPG-to-ECG reconstruction transformer for a digital biomarker of cardiovascular disease detection. In: Proc IEEE/CVF Winter Conf Appl Comput Vis, pp 1991–1999
Mundu A, Singh SK, Dubey SR (2024) ETransCap: efficient transformer for image captioning. Appl Intell 54(21):10748–10762
Nabi M, Pachauri R, Ahmad S, Varshney K, Goel P, Jain A (2023) Visual image captioning through transformer. Int J Res Appl Sci Eng Technol 11(12):2047–2050
Navalkar MDG, Wankhade NR (2024) Smart Caption: Intelligent image description using transformer decoder. Int J Adv Res Sci Commun Technol, pp 144–150
Ning X (2024) Linear Attention is Enough in Spatial-Temporal Forecasting. arXiv preprint arXiv:2408.09158
Ondeng O, Ouma H, Akuon P (2023) A review of transformer-based approaches for image captioning. Appl Sci 13(19):11103
Papineni K, Roukos S, Ward T, Zhu WJ (2002) Bleu: a method for automatic evaluation of machine translation. In: Proc 40th Annu Meet Assoc Comput Linguist, pp 311–318
Pardede J, Fandi (2024) Image captioning using Transformer with image feature extraction by Xception and Inception-v3. Kursor 12(3):135–146
Sun D, Bao Y, Cao X (2024) A Lightweight Transformer for Remote Sensing Image Change Captioning. arXiv preprint arXiv:2405.06598
Tang J (2024) Best RNN+ CNN combination in image captioning. In: Proc 5th Int Conf Comput Sci Manag Technol
Xiong Y (2021) Nyströmformer: A nyström-based algorithm for approximating self-attention. In: Proc AAAI Conf Artif Intell 35
Yang X, Liu Y, Wang X (2022) Reformer: The relational transformer for image captioning. In: Proc 30th ACM Int Conf Multimedia, pp 5398–5406
Young JC, Arthur R, Williams HTP (2023) CIDER: Context sensitive sentiment analysis for short-form text. arXiv preprint arXiv:2307.07864
Zhang P et al (2021) Multi-scale vision longformer: A new vision transformer for high-resolution image encoding. In: Proc IEEE/CVF Int Conf Comput Vis, pp 2998–3008

Engineering Innovation Meets Strategic Excellence: Navigating the Technological Revolution

Preeti Malik, Varsha Mittal, and Kamika Chaudhary

Abstract The dynamic fusion of engineering innovation and strategic excellence serves as a bridge for industrial technological revolution. The disconnect among technologists and management in terms of targets and metrics of success often leads to poor decision-making and business growth. With the rapid industrial advancement in automation, artificial intelligence and digital transformations, organizations must integrate cutting-edge engineering solutions with forward-thinking strategies to maintain a competitive edge. The chapter outlines the major factors responsible for reshaping traditional business models, from the rise of Industry 4.0 to the growing importance of data-driven decision-making. It also highlights the challenges organizations face in adapting to these changes, including the need for agile infrastructure, a skilled workforce, and effective leadership. By framing technological transformation as both a catalyst for innovation and a source of disruption, the chapter sets the foundation for understanding how businesses must evolve to remain competitive and thrive in an increasingly interconnected and rapidly changing world.

Keywords Engineering · Artificial intelligence · Management · Project · Technology · Internet-of-things · Technology-driven environment

> Engineers are good at solving technical problems. However, they're often not as strong at leveraging data science techniques to make evidence-based decisions, or leading and inspiring diverse teams with empathy and inclusiveness.—C. Fred Higgs

P. Malik (✉)
Uttar Pradesh State Institute of Forensic Science, Lucknow, India
e-mail: preetishivach2009@gmail.com

V. Mittal
Graphic Era University, Dehradun, India

K. Chaudhary
M. B. Govt. P. G. College, Haldwani, India

M. Ram et al. (eds.), *Technological Advancements in Engineering and Management Sciences*, Management and Industrial Engineering,
https://doi.org/10.1007/978-3-032-04606-2_12

1 Introduction

The technological innovations are evolving at a high pace, together with transforming other industries across every area (George and George 2024). Innovations in everything from biotechnology and nanotechnology to robots and Artificial Intelligence (AI) is allowing previously unheard-of capabilities and upending long-standing economic structures. These technologies help businesses to grow. Like AI improves the business by providing data-driven insights for better decision-making and optimizing the use of resources, whether it is man-power or some other resource.

It would not be an exaggeration to say that AI may be the most disruptive technology available. AI is about making computers intelligent like humans. Machine Learning (ML) is a subset of AI, which makes systems to learn automatically and improve their performance without explicit programming. ML algorithms basically use pattern detection and build statistical models for performing specific tasks like natural language processing, image recognition, and predictive analysis. Even the industries are using AI for quality control, demand Prediction, dynamic pricing, etc. The AI industry is estimated to reach a size of US$243.70 billion in 2025, with a projected annual growth rate (CAGR 2025–2030) of 27.67%, reaching a market volume of US$826.70 billion by 2030. Figure 1 shows the growth of the market for various technologies. Evolution in *robotics* is making the systems more sensible, skilled, and intelligent for automating tasks and improving collaborative work with humans. Robotics is emerging as smart helping technology, which is integrating itself into human space. The manufacturing industry is transformed with the growth of collaborative robotics by interacting with the workers (Pereira et al. 2013; Zafar et al. 2024). Other industries like oil and gas, security, and agriculture are also getting

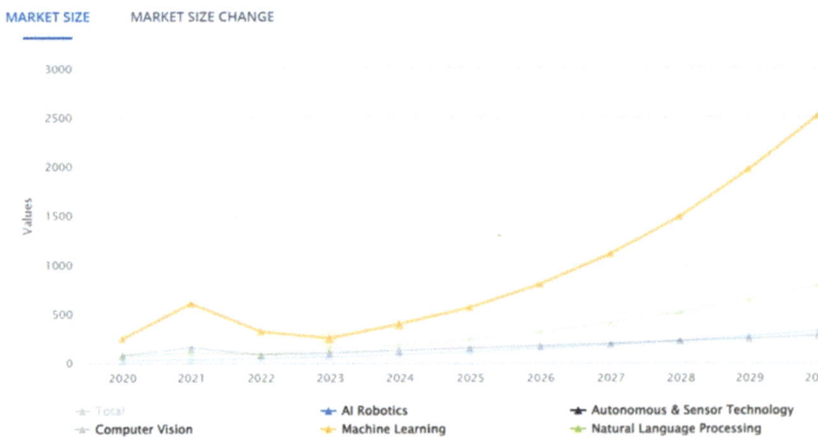

Fig. 1 Market for various technologies https://www.statista.com/outlook/tmo/artificial-intelligence/worldwide#market-size

benefits from robotics in terms of inspection and surveillance. Talking about the market share, the intelligent robotics market is forecast to surpass $32 billion by 2027 {source: https://www.fortunebusinessinsights.com/smart-robot-market-104556}.

Internet of Things (IoT) is another technological evolution which refers to the networks of interconnected devices that work collaboratively (Hozdić 2015). Data is gathered from various devices and processes. IoT empowers organizations in monitoring the equipment from a distance, tracking the assets, and optimizing the system in real-time. By 2025, it is anticipated that there will be over 75 billion IoT devices in use, forming a massive networked ecosystem that will provide enormous amounts of useful data. By 2028, the worldwide IoT industry is expected to grow to around $1.6 trillion.

Blockchain is also becoming the center of attraction in the technology world. Blockchain is a decentralized ledger system that promotes efficiency and trust by providing safe, transparent record-keeping for supply chain management, transactions, digital identity verification, and smart contracts (Nakamoto 2008; Sharabati and Jreisat 2024).

Augmented Reality (AR) and Virtual Reality (VR) refers to creating immersive experiences by overlaying digital information onto the real world (AR) or simulating entirely virtual environments (VR). It is basically used for customer engagement, the entertainment sector, and training purposes. Quantum Computing is all about using quantum mechanisms to solve complex problems much quicker than traditional computers. It is now used in the discovery of drugs, materials science, and cryptography. Ultra-low latency and high bandwidth communication are being made possible by 5G networks for real-time applications such as sophisticated industrial automation, remote surgery, and driverless cars.

1.1 Bridging Engineering and Management Disciplines

There is a disconnect between what engineers are producing and what management perceives and comprehends. Poor business decision-making, career stagnation, and unhappiness can be the result of the work culture gap between engineers and management (Buckbee 2010). This distance encompasses not only a communication gap, but also a gap in understanding and expectations. The two groups have different goals, different metrics for success, and different results in mind. Because it enables engineers to not only design and develop technical solutions but also comprehend the business context, project constraints, financial implications, and stakeholder needs, bridging the disciplines of engineering and management is essential. This leads to better decision-making, more successful product commercialization in the market, and more effective project execution.

In an organization or business, engineers may make decisions for delivering the product on time by keeping it within budget and efficiently utilizing the finance and resource management. The communication gap between stakeholders and the technical team can also be reduced by filling this gap, which makes the project goal clearer. Even a person with knowledge in both engineering and management can lead a cross-functional team effectively by motivating and supervising them toward project goals. Market awareness is another field where this engagement of engineering and management may give better results. A person can realize the dynamics of the market and needs of the customer, which in turn can be applied to designing the products that are sound both technically and commercially.

2 Artificial Intelligence Driven Innovation

It is becoming essential for engineering managers to cope with the growth and acceptance of AI in managing both the human and AI resources. A flexible approach for effective use of AI to improve productivity and innovation in a business is needed. It is a crucial step to develop approaches or policies to integrate an AI tool into the existing workflow of a business or organization. Such a type of integration demands a thorough understanding of the pros and cons of integrating AI tools. Hence, engineering managers must estimate where this integration can add more value to the output, whether it is automation, optimization, or anything. Still, this integration or technological shift doesn't lessen the prominence of foundational engineering practices. Various studies like (Rashid and Kausik 2024; Li et al. 2017; Sarker 2022; Anantrasirichai and Bull 2022; Javaid et al. 2022) conducted review research on the use of AI in intelligent manufacturing, highlighting current applications and developing an assessment of the "Internet plus AI" era. Actually, it is difficult to balance these traditional practices and the challenges of AI. An engineering manager is required to keep focus on the dynamic forces of a team, quality control , and management of the project, along with adjusting to the complexities arose from the navigation toward AI technologies. Use of AI-driven technologies sometimes exceed human intelligence at least up to some degree. This section is dedicated to areas where AI can be applied in a business or organization.

2.1 Customer Service and Support

Most of the organizations make use of chatbots as customer service to interact with people, and this is achieved with the help of Natural Language Processing (NLP). Tasks like tracking orders or answering frequently asked questions are also being done with the help of AI. Such applications should work like a supplement to a human rather than a replacement. It helps in freeing up the human workforce to give attention to a more complex or high-valued task.

2.2 Marketing and Sales

According to McKinsey, AI applications in marketing and sales could reach a value of $2.6 trillion. Hidden truth or uncovered patterns could be provided by AI-driven analysis techniques, which in turn make the marketing more successful. These applications can offer smooth interaction, enabling users to move between channels without losing their work. AI is also helping in generating marketing strategies mainly based on the behavior of customers and market trends (Haleem et al. 2022; Capatina et al. 2020; Frank 2021; Devang et al. 2019).

2.3 Supply Chains and Logistics

Another sector that benefits from the use of AI is a supply chain. McKinsey also highlighted that the value of AI applications is expected to reach $2 trillion. Demand prediction, inventory optimization, and improving supplier relationship are one area where AI is being used extensively. Although AI is not a new topic or academic discipline (Toorajipour et al. 2021; Huin et al. 2003), technological advancements have only recently demonstrated that AI has a wide range of applications (Min 2015), garnering attention for its ability to adapt processes in a wide range of diverse fields, including supply chain management (SCM). AI technology is becoming a competitive advantage, whereas other fields of information technology are being relegated to a position of competitive necessity (Thow-Yick and Huu-Phuong 1990).

2.4 Financial Services

Some of the highly regulated industries, like the financial sector, are also benefiting from the use of AI-driven technologies. AI is being used in the detection of anomalies for fraud detection or cyber scams. AI-based tools can also be used to automate the aspects of compliance to improve productivity and reduce human errors.

2.5 Human Resources

Researchers are currently concentrating on developing strategic HRM practices supported by AI technology has become widely accepted and received a lot of attention (Nawaz et al. 2024; Wang et al. 2020). NLP can be used in creating HR related contents. The data analysis feature of AI can be used to determine the most effective recruiting channel. Time and labor-intensive task can also be achieved with the help of AI-based techniques. AI also helps in assessing the employee performance and predicting the future talent trend in building the organization's future-proof workforce.

2.6 *IT and CyberSecurity*

AI-based security tools can implement automated penetration testing to identify loopholes in the system and find security defects before the applications are deployed. Monitoring the network traffic can also help in the early detection of some kind of unauthorized access. Flagging the potential scam emails can also be done by using AI tools. Das and Sandhane in (Das and Sandhane 2021) study assesses the potential for increasing cybersecurity capabilities by strengthening the defense mechanism and gives a succinct review of AI implementations of various cybersecurity employing artificial technologies. AI applications can shorten time to market and iteration once a company has established a high demand for a product. AI tools can create a prototype or mock-up in hours rather than weeks, and they can scan and analyze customer sentiment data from multiple sources to create faster and more accurate feedback loops.

3 The Interdisciplinary Approach

The engineering industry is reforming itself under the age of technological transformation at a very fast pace. Other than AI, there are other technologies that are being used in the field of Engineering. Engineering practices are being revolutionized by technology, whether it is design, construction, simulation, or maintenance, every field is being affected by this revolution. The conventional, time-consuming manual calculations, drawing, physical testing, and model construction methods have been progressively replaced by computer-based modeling and simulation since the 1960s for many design activities (Dodgson et al. 2007). In the construction industry, creating models of buildings is an application of technology. These models are called Building Information Model (BIM), which represents the physical and functional features of a building that is to be built by engineers, architects, and also the contractors. A three-dimensional model of the building is created with the help of technology, and this model is further used for design and analysis throughout the lifecycle of the building. Use of this BIM enriches the collaboration among all the stakeholders and also ensures that up-to-date information is being shared by the stakeholders. Even the potential danger can be identified before the construction of the building, which in turn reduces the time and cost of building.

Another use of technology in the manufacturing industry is 3D printing, which is also called additive manufacturing. 3D printing helps in creating complex components that are difficult to build using conventional methods of manufacturing.

Reduction of material waste is one of the major advantages of 3D printing. In traditional methods of manufacturing, a lot of material is wasted, which also cannot be reused, whereas 3D printing only uses essential materials. This feature of 3D printing presents it as a sustainable option. AR, VR and IoT have already been discussed in the previous section. The path toward sustainable development is significantly affected by engineering techniques and methodologies (Glavič 2022). Sustainable engineering is nowadays of great importance to ensure that new technologies do not affect the environment around us. Solar panels and wind turbines, which are forms of renewable energy, are being combined into building projects to lessen dependence on fossil fuels. Engineers are focusing on a design that causes minimal damage to the environment. Engineers use some kind of life cycle assessment tool to assess the environmental impact of a project from the beginning to the end of the life cycle. Based on this assessment, the engineers make decisions that align with sustainability goals.

3.1 Collaboration Between Technical Teams and Management Professionals

In order to promote innovation, increase efficiency, and improve the overall performance of your enterprise app development firm, cooperation and communication between the IT and business teams are essential (Buckbee 2010). In addition, there are several other advantages and justifications for embracing a dynamic workplace culture in order to facilitate efficient communication and cooperation between the various teams engaged in the creation of bespoke business software. A collaborative working culture promotes faster completion of a task and optimum utilization of resources. On the other hand, rich communication benefits in arranging work, reducing delays, and making fast decisions, which in turn lead to improved productivity, better efficiency, and healthier outcomes. It also brings miscellaneous viewpoints together that support organizations to effortlessly solve difficult problems with better agility. More state-of-the-art and effective solutions are the result of such diversity of thought. Collaboration also built trust and bonds among the members of the team. It helps in developing a mutual empathy among them. Team members can share their plus and minus points and inspire each other to make efforts with passion, and support each other.

Success in the fast-paced corporate environment of today depends on efficiency, creativity, and agility. "Collaboration is the new competitive advantage," as the saying goes. Cross-departmental collaboration is essential to achieve innovative approaches for operational effectiveness and organizational success. Cross-departmental cooperation is explained as the process of individuals or teams from various departments working together to accomplish a shared objective of the organization. This type of collaboration, where knowledge, resources, and experience is shared among different departments, will enhance the overall performance of the organization and will help to accomplish greater business objectives. The motivation for cross-departmental collaboration is the thrust for innovation, solving complicated issues, and improving the performance and productivity of the organization. An ultimate level of customer satisfaction and productivity can be achieved by promoting the concept of shared ownership and dismantling organizational divisions. Here's how:

- Establish a Learning Environment: There is a need to create an environment that focuses on updating and transferring knowledge by conducting events like hackathons that build the problem-solving capabilities using innovative and creative methods, promote cooperation, and cultivate a building habit.
- Embrace Outcome Over Activities: To apply the approaches that change the perspective and prioritize the result.
- Collaboration versus Competition: It is required to motivate for the development of an environment where cooperation supersedes rivalry, the team should have the collaboration values so that they can come together to work for organization-wide advantages rather than single-department initiatives.

In the collaborative environment where knowledge exchange is practiced, it results in enhanced employee skill development. When people from different fields come closer to work together, they learn new skills, acquire knowledge that will enhance their performance and productivity. This collaboration always increases the efficiency of the employees, which leads to timely task completion (Sayed et al. 2022). Secondly, different perspectives are brought on the same platform when various departments come together. This makes the decision-making process more well-rounded and knowledgeable. Using this multiplicity of perspectives, managers are able to identify the probable risks as well as new opportunities and can take more strategic decisions that can improve the company's overarching objectives.

The convergence of different perspectives usually results in innovation. When people from different departments with different skills and different problem-solving methods work together, it results in better and efficient solutions. New services and creative ideas may lead to new business opportunities. Also, cross-departmental collaboration fostered the culture of trust in an organization. When different people from different departments work together, they can understand the roles and responsibilities of each other, which will build the feeling of commonness and trust among them and will help to minimize misunderstandings. It will improve the working environment, and the organization can cultivate a more transparent and healthy work culture.

4 Impact on Project Management

To manage the various aspects of project management, like monitoring project performance, resource allocation, and streamlining the project activities use the AI. The use of AI in project management helped the managers to efficiently plan and analyze

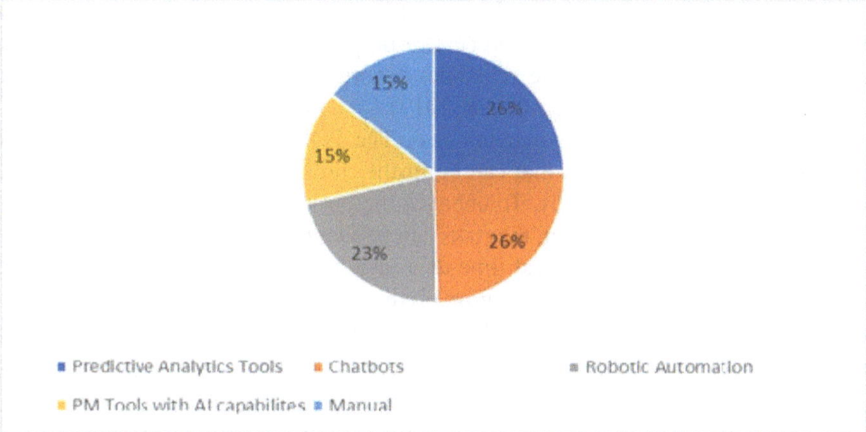

Fig. 2 AI usage in project management (*Source* http://Microsoft PowerPoint - 20201023 IPMA PwC AI Impact in PM (NEW RED).pptx)

the projects that will increase the productivity of one hand and decrease the cost simultaneously. Use of AI in a project management is increasing day by day, so that the managers might have the real-time data and insights (Vaswani 2017). AI can also automate the different monotonous processes of the project management (Kim et al. 2022). Since AI can deliver real-time insights and make predictions that can assist the project managers to diagnose the problem in time and take the suitable action plan to deal with it. Figure 2 shows the usage of AI in project management according to a survey report, which was published in 2020 by International Project Management Association, in association with PWC.

4.1 Streamlining the Project Scheduling

Project scheduling is still one of the tedious tasks in project management. The challenges of managing changing schedules, forecasting potential delays, and coordinating global resources require a lot of experience. The evolution of generative AI in the scheduling process of a project has revolutionized project management and is a boon to the managers. It provides an innovative cutting-edge to the project scheduling (Alexandropoulos et al. 2019).

The use of AI in scheduling provides a method for predictive risk identification and optimized resource management. However, generative AI is new to this field, and a lot of scope is still there where its usage can improve the efficiency of the system. In the case of complex projects, generative AI offers a competitive advantage as scheduling plans can be modified dynamically , considering the real-time insights and scenarios. This automates the process of project scheduling, simulations of the scenarios, and adjustment of the schedules according to the dynamically changed requirements and situations (Dash et al. 2019).

Further AI has some other benefits also. The PDFs of contracts can be easily transformed into Gantt charts that can be further broken into small schedules, and action can be taken accordingly. To enhance the project flexibility and to meet the deadline for project completion, the risks can be identified at the earlier stage since AI has the capacity to deliver real-time data insights. Moreover, since the possible risks at each stage of the project can be predicted using generative AI, it helps to estimate budgets for each activity. Hence, more accurate cost estimation can be done. The summarized report of every ongoing activity can be made available on the dashboards, so it becomes quite easy for the managers to track the status of the project completion. AI is able to work on real-time data, so the project performance can be monitored regularly. In time, the warning delay messages can be sent to the project team so that the project completion can meet the deadline, keeping the constraint of budget (Engelmann et al. 2020).

The challenges associated with the use of AI in scheduling includes: data quality and human insights. Since AI works on data, maintaining data quality is a challenging task. Hence, ensuring the correctness and relevance of data is significant. However, AI can automate the task of project scheduling efficiently, but human intelligence is required to interpret and make decisions in some unforeseen situations.

4.2 Decision-Making in Resource Allocation

The incorporation of AI into project management has given a new dimension to the different aspects, such as resource allocation, risk mitigation, and decision-making with reference to the project. Because of the data analysis and real-time amendments, the project managers are capable of handling unforeseen situations (Schwung et al. 2020). The resource management is done in a more effective and efficient manner; moreover, changing day-to-day requirements can also be handled. Different machine learning algorithms are used for allocation of resources like staff, equipment, and finances. This will increase the productivity, do the cost cutting, and result in better resource utilization.

Balancing the triple constraint of time, money, and scope of project management is quite difficult while doing the manual resource allocation. It becomes more critical with the growing size of the project as it may include several dependencies. Delay in one task will affect the other dependent operations. The machine learning algorithms used for resource allocation use both the current and past data, resulting in more accurate predictions about the future resource requirement trend. AI in resource allocation is used in the following ways:

- Data Analysis: The different algorithms used in AI process and analyze the vast amount of data, including historical data of past projects, ongoing projects, and the current project metrics. It provides a complete perspective of the resource requirements of the project, which is otherwise difficult to do manually. The results of these algorithms can assist the project managers to predict the requirements more accurately and proactively allocate resources
- Predictive Modeling: The predictive models can be designed that use both the internal variables, like progress measurements, historical performance data, and external environmental variables, to automate the process of resource allocation. These models can help the managers to anticipate the scenario of either a bottleneck or a surplus in predicting the future resource requirements trends. AI when used with reinforcement learning, where the systems learn by real-time feedback always improve the prediction and hence help to design a more dynamic system for resource allocation, where traditional methods would find it difficult to keep up.
- Real-time adjustments: One of the strongest contributions of the use of AI in resource allocation is its ability to facilitate real-time resource allocation. AI systems can collaborate on the data from different sources like the software used for project management, sensors used in Internet of Things (IoT) and the communication platforms. They can track the progress and utilization of resources more precisely. Also, by evaluating the project completion rates, interruptions, and resource consumption rates, the AI systems can reallocate the resources

The resource allocation done through AI powered systems ensure the maximum utilization of resources, reducing the idle time and increasing the throughput of the system. Automation of job scheduling and resource allocation reduces the burden on project managers, which increases their productivity in other strategic decisions. AI systems can predict the situation of overallocation and underutilization of resources. It will prevent the expensive last-minute resource requirement to meet the project deadline. Since AI systems can predict the possible shortages or bottlenecks in resources before the situation actually appears. Project managers can maintain project resilience by planning to handle situations and avoid project delays. The decisions made in AI powered resource allocation are based on the insights and analysis of a vast amount of data. Therefore, the resource distribution is done with objectivity and knowledge and reduces the probability of human error or preconceptions (Zappone et al. 2019). Automated resource allocation ensures optimized resource allocation throughout the project lifecycle. AI-powered resource management makes it easier to complete projects on time, under budget, and with high quality.

5 Case Studies and Real-World Applications

5.1 Apple: Integrating Design, Technology, and Management for Consumer-Centric Innovation

The most seamless confluence of technology and management is adopted by Apple Inc. Its approach to innovation is the prime example of the way technology and management converge seamlessly. The company focused on creating a robust ecosystem of products and services from the iPhone to the Apple Watch. Apple has adopted a combination of cutting-edge technology along with the management strategies that focus on user experience, aesthetics, and simplicity. The success of Apple stands on two strong pillars: one is its ability to produce strong hardware, and the second is the way it manages its supply chain, distribution networks, and software development. It focuses on providing a holistic and integrated experience for its customers.

The innovations are welcomed in Apple, and to foster an innovative culture, management plays a crucial role. Under the leadership of Steve Jobs and later Tim Cook, Apple has maintained an unwavering focus on design and quality, which has become synonymous with its brand. The philosophies behind both leaders focus on creating a product that is not just on par in functionality but also intuitive and beautiful. For instance, the company has tightly controlled its software and hardware integration, which allows for a seamless experience across devices.

Apple also put quite a good amount of focus on its product management. The ability to pursue a consistent and clear strategy and swiftly respond to new consumer market trends positioned Apple as a leader in the navigation aids market. This convergence of technology and management has also extended to its retail operations, where Apple's stores are designed to be not only places to buy products but also hubs for customer engagement and brand loyalty. By doing so, Apple has created a technology-led ecosystem that brings together hardware, software, and service management, all aimed toward delivering exceptional value and experience to its customers.

5.2 Tesla: Disrupting the Auto Industry with Technology-Driven Management Innovation

Tesla is a perfect example of a business using strategic management, combined with technological innovation, to disrupt an entire industry. It has advanced the automotive industry by introducing electric vehicles (EVs) with the advantage of cutting-edge battery technology and sophisticated autonomous driving systems. Tesla has redefined business management in the automotive industry. Its management strategy has been centered around a long-term vision that aims to achieve the sustainability and to use renewable energy. It focuses on making technology and business management go hand in hand mutually (Lopez Conde and Twinn 2019).

The focus on innovation has become one of the most important ingredients of the company's management style, especially under the leadership of CEO Elon Musk. Musk has established an environment where people pursue technological breakthroughs, such as full self-driving (FSD) software and the Gigafactory, as a large part of achieving greater strategic goals. The Gigafactory makes EVs more affordable and sustainable by reducing the cost of batteries and scaling up production. Tesla has gained a competitive edge over traditional automakers with the help of its strategic management decisions of vertically integrating the supply chain (Ingle and Phute 2016).

To optimize its manufacturing process, Tesla has utilized data-driven management techniques. The company's over-the-air software updates, which allow for continual improvement of its vehicles' performance and safety features, are a testament to how technology and management work together. Bringing together the machine learning algorithms along with real-time data, it has improved its car efficiency and overall user experience. Tesla has effectively linked technology with administration to crush typical car companies and redefine what cars can do in the automobile sector (George et al. 2023).

5.3 Amazon: Cloud Computing and Supply Chain Management Synergy

Amazon's business model exemplifies how technology and management can converge to create new opportunities for growth and market leadership. Amazon has come up as a global e-commerce and cloud computing powerhouse from an online bookstore by adopting cutting-edge cloud computing technologies along with world-class supply chain management. Key strategic decisions, such as the creation of Amazon Web Services (AWS) and innovative logistics systems, have been crucial to its achievements.

The launch of AWS in 2006 has revolutionized cloud computing services by expanding the infrastructure of the business at a much lesser in cost than traditional IT infrastructure. AWS has allowed Amazon to make use of its existing technology to create a new revenue system and become influential in the tech space. This strategic move of the management has proved to be fruitful for Amazon, not just in the sense of growing demand for cloud services from other sectors, but also in establishing it as a king in the emerging technology sector.

At the same time, Amazon has also focused on the logistics department and changed the fate of supply chain management. It is able to develop a supply chain that is quicker, efficient, and highly in sync with the customer demands. All this has become possible due to the incorporation of technologies like robotics, machine learning, and data analytics techniques. Amazon has been able to maintain the best experience for customers due to its AI application for demand forecasting, inventory optimization and personalized recommendations. The technology and management strategies have come together for Amazon and paved the way for its commitment toward next-day or even same-day delivery, making it as a leader in the e-commerce industry.

5.4 Siemens: Industrial IoT and Smart Manufacturing

In the manufacturing sector, Siemens has established itself as a leader in industrial automation and digitalization, setting a global standard for innovation. Through its adoption of the Industrial Internet of Things (IIoT), Siemens has transformed the way manufacturing processes are automated and streamlined. By integrating cutting-edge technologies such as sensors, AI, and cloud computing into its products, the company has developed solutions that enhance efficiency, minimize downtime, and boost overall productivity. Siemens's has developed "Digital Factory" as its flagship concept to present and showcase how technology and management coming together can influence the manufacturing environment to become smarter and more efficient. This concept of digital factory has helped in digitizing the entire production lifecycle, beginning with design and then to production and post-production. It has leveraged on technologies like digital twins, real-time data analysis, and cloud-based platforms. The approach of the digital factory proves to be beneficial in terms of bringing flexibility in the business so that they can adapt according to the changing market quickly.

Siemens has been able to move ahead in Industry 4.0 revolution by offering its expertise as a digitalization partner in to other competitors in the market. It has been able to provide tailored solutions by using IoT technologies in its product offerings across a wide range of industries, including automotive, aerospace, and energy. Siemens has established itself as both a technology leader and also a strong ally for the companies that want to stay competitive in a technology-based market (Annanth et al. 2021).

5.5 GE Healthcare: Merging Medical Technology with Data-Driven Management

In the healthcare sector, blending technology with effective management has brought innovation in this field. GE Healthcare has proved to be a prime example that has benefited from this collaboration of technology and management. It has been able to develop sophisticated diagnostic solutions by integrating cutting-edge medical imaging tools with advanced data analytics and AI technology. These technology-driven solutions help to provide doctors with deeper insights and more accurate diagnoses. The company has also been able to enhance patient care and streamline operations by embedding data management capabilities into its healthcare devices (George and George 2023).

The management approach of GE Healthcare concentrates on bringing together healthcare professionals with data scientists and engineers in order to create products that can meet the specific needs of the industry. GE has been able to make use of AI in medical imaging, specifically in systems like MRI and CT scanners. The embedding of AI algorithms with imaging tools have resulted in reducing the amount of time needed by medical practitioners to analyze and detect abnormalities in the scans. This has ultimately enabled the healthcare facilities to run speedily and smoothly.

GE Healthcare has shifted its focus to patient-centric solutions. This has resulted in the development of remote monitoring support systems and enables the sharing of data among patients and their healthcare providers. It has improved patient outcomes, reduced costs, and boosted operational efficiency by incorporating technologies like cloud platforms and IoT devices. This fusion of advanced medical technologies with data-driven management practices in GE Healthcare is reshaping the healthcare sector by providing better care to patients around the world.

6 Impact of AI on Real-World Business Projects

The impact of AI on real-world business projects is huge as well as multifaceted. Its impression starts from enhancing the customer experiences to streamlining operations, improving decision-making, and also focuses on providing innovations. AI is working at the forefront of business transformations, and as it continues to evolve, its part in shaping the future of business will also grow. AI offers new opportunities as well as throws the challenges related to ethics, security, and workforce adaptation for organizations across all industries. AI automation system technology is playing a vital role in transforming the business productivity, product consistency, and optimized resource utilization. AI embodies the forefront of the transition to Industry 5.0 as it is concentrated on the collaboration between humans and AI Systems. AI has emerged as a vital technology in recent years, finding applications across various industries. It plays a crucial role in many fields, transforming industries and tackling significant global issues (Biswal 18).

AI has made a great impact on changing the game for real-world business projects. The industries now follow the AI driven path for problem-solving, innovations, and strategic executions. Now businesses are able to achieve unprecedented growth rates as the complex tasks, decision-making, and strategic execution are supported by AI. Its impact is not just limited to a few but spans across various sectors ranging from manufacturing to retail, healthcare to finance, and beyond. AI's capability to provide data-driven insights has become one of the most significant ways of influencing business projects. It allows businesses to analyze a huge amount of customer data so that their future behavior can be predicted. On the basis of predictions, the industries make their customer-specific target advertisements and personalize their experiences as well. Big giants like Amazon and Netflix use AI algorithms to curate recommendations on the basis of user preferences and their past interactions on the websites, which in turn ensures customer engagement and loyalty. This ability to provide customer-specific personalized experience has made the business invest in AI—driven customer relationship management systems.

Industries like manufacturing and logistics have faced a great advancement in automation with the collaboration of AI in business operations. With the combination of robotics and AI powered systems precision and efficiency in redundant tasks have been improved. This has also limited the scope of human error. These AI-powered robots are performing the assembly line tasks efficiently, resulting in an efficient production cycle and reduced costs. AI has also enabled predictive maintenance, which can detect malfunctions in the system before their occurrence, minimize downtime, and ultimately prevent costly repairs.

AI is also reshaping the finance sector landscape. Risk management and fraud detection are identified by using machine learning algorithms. Various suspicious activities are identified by analyzing a large set of data. Various organizations like JPMorgan Chase and Bank of America ensure greater security for their customer by continuously monitoring the transactions and performing the assessment of risks. AI also enables a personalized financial product and services experience for customers.

7 Future Trends in Technology and Management

The future of AI across various industries is undoubtedly bright and shiny with endless possibilities of innovations and progress. As it continues to advance, it promises to be transformative and reshape the way businesses operate, the entire sector functions, and compete. The healthcare industry will observe how AI plays a pivotal role in each and every sector, beginning with diagnostics, personalized medicines, and drug discovery. AI-rich algorithms will analyze the patients' medical images and provide much better and accurate diagnosis of diseases in their early stage only. Furthermore, AI will also provide personalized care plans based on individual-specific genetic designs. It will also expedite the process of bringing the lifesaving drugs to market by identifying the promising compounds.

AIin the financial sector will continue to revolutionize areas such as risk management, fraud detection, and personalized financial services. Various machine learning algorithms will become more advanced in analyzing the market trends, predicting price movements, and providing real-time information for traders and investors. AI will also focus on safeguarding the sensitive financial data by enhancing cyber security, fraud mitigation, and suspicious actions identifications (Pallathadka et al. 2023). In the manufacturing industry, AI will make the entire industry undergo a dramatic transformation as it powers the next phase of smart manufacturing. The factories will become completely autonomous with the adoption of AI, robotics, and IOT. The work that AI will facilitate will include optimization of production schedules, equipment health monitoring, customization, consistent quality control, and what not. The presence of AI and its technologies will improve both efficiency and customer satisfaction (Arinez et al. 2020).

In retail and e-commerce, the future role of AI will be centered as both these industries will see remarkable advancement through AI. The experience of the customer, management of inventories, supply chain optimization, and AI-powered chatbots, 24/7 customer support, and guidance at purchases will be transformed by AI. It will enable predictive analytics so that demands can be forecasted more accurately, and also reduce the stockouts and excess inventories. The delivery timelines will be improved with AI presence (Vanneschi et al. 2018).

The adoption of AI in the transportation and logistics industry will work remarkably well, particularly in the development of autonomous vehicles and smart logistics networks. The goods will be transported with the self-driving cars, trucks, and drones that will reduce human interventions and increase operational efficiency. In urban transportation, AI will facilitate the development of smart cities, where traffic flows are optimized through predictive algorithms, and public transportation systems are more efficient and responsive to demand.

Looking ahead, the adoption of AI in creative industries like media, entertainment, and design will also expand. AI algorithms are poised to play a significant role in content creation, whether it's generating music, writing, or producing visual art. With AI-powered tools like deepfake technology and generative design, new forms of creative expression will emerge, although these advancements will also prompt ethical discussions around issues of authenticity and ownership. In the entertainment sector, AI will drive more personalized content recommendations, changing the way consumers discover and interact with media.

Looking ahead, AI's role across industries is set to evolve from isolated applications to an integral part of overall business strategy. As AI continues to advance, and its capabilities will increasingly mirrors human cognitive functions, making it indispensable for tasks involving reasoning, creativity, and even emotional intelligence. However, the rise of AI also brings a host of challenges, particularly in areas such as ethics, privacy, and job displacement. Industries will need to address these concerns through responsible AI development, ensuring that the technology's growth is beneficial to society while minimizing potential risks. The impact of AI will continue to expand, transforming industries in ways that are just beginning to take shape.

8 The Evolving Role of Management in Technology-Driven Environments

With the advent of new technologies such as AI, Big Data, cloud computing, and automation, the role of management is undergoing a profound transformation. Now the traditional management functions are being redefined. They have to be no longer just decision-makers but now required to be strategic visionaries, supporters, and facilitators of innovations and the first to adapt to the technological shift. They have to be the leaders who can guide the organizations through technological disruptions.

8.1 Leadership in Digital Transformation

The role of management has shifted toward identifying and implementing new technologies and also fostering the culture of innovation in organization. Digital transformation is not just related to the adoption of digital tools rather it's a fundamental strategic initiative that impacts all aspects of an organization. The responsibility of the management now stretched toward guiding the organization through technological disruptions, and making the employees understand the vision behind these changes. Whether it's moving to the cloud, implementing AI-driven tools, or embracing automation, managers must be proactive in setting the tone for change, helping the company stay agile, and ensuring technology is leveraged for competitive advantage.

8.2 Data-Driven Decision-Making

Data-driven insights are critical assets and help managers to make more informed and strategic decisions. In a technology-driven environment, the rise of big data and AI tools makes the manager task to use a complex dataset in order to understand the market trends, behavior of customers, and operational efficiencies. It has become important for managers to be proficient in analytics tools and to collaborate with a data scientist in order to extract the meaningful insights and make good business decisions. For instance, Amazon has revolutionized retail by using data analytics to forecast demand, personalize recommendations, and optimize supply chains. Managers at Amazon must navigate vast datasets, using algorithms and AI to continuously refine the business model and improve the customer experience.

8.3 Managing Innovation and Agility

With the rapid pace of technological change, fostering an environment of innovation and agility has become essential for businesses. Earlier agile methodologies were limited to software development only, but now they have spread across a number of business functions. This change has made the management challenge their strategies and form cross-functional teams, which can respond positively and rapidly to technological advancements. With the fastest-growing technological landscape, the ability to iterate, experiment with new ideas, and implement solutions efficiently has become critical. It has become necessary for the managers to focus on innovation, along with the need for balancing the speed and maintaining quality and sustainability. As an example, Tesla is the company that has proved itself to be a perfectionist in the art of innovation, regularly rolling out new features for their electric vehicles through software updates, all while continuously improving the performance of their products.

Additionally, the manager's responsibility has now increased to create an environment that values experimentation and takes failure as an opportunity to learn. This kind of environment and mindset allows the teams to take risks and explore innovative solutions without fearing mistakes and the cost they incur.

Earlier, the role of management was confined to overseeing the daily operations, but now it has evolved drastically into visionary leadership. Managers today must not only harness technology to drive growth but also foster collaboration and ensure ethical and responsible use of technology. As technological advancements continue to reshape industries, managers must stay agile, continuously update their skills, and remain ahead of emerging trends. The ability to lead through these rapid changes will ultimately determine the future success and resilience of businesses.

9 Ethical Considerations

To use the technology to its fullest, there must be some kind of ethical concern related to its acceptance. Ethics plays a major role in accepting technologies without any accidental costs and prejudice. Ensuring accountability and building confidence between companies, individuals, and governments makes it easier to employ technology effectively. Even though quickly evolving technology presents a lot of ethical challenges, businesses must ensure that personal data is protected and managed ethically. Though there could be more, these are some of the most important moral challenges.

9.1 Personal Data Misuse

Misuse of personal information, i.e., privacy breach, grows as a major ethical problem. This huge amount of personal information is collected by corporations via websites like social platforms, e-commerce websites, etc. (Mouw et al. 2023). Businesses gather the data in order to tailor our experiences, learn what kinds of goods and materials we are interested in, or just to increase their customer. Such access to our personal information may also be observed as a privacy violation and may result in adverse situations such as data breaches and cyberattacks. These corporations may share our personal information with third-party corporations and use it inappropriately for targeted marketing.

9.2 Dissemination of False Information

The availability of real-time updates and information due to continuous internet access increases the possibility of misusing or simply spreading information without fact-checking. This kind of erroneous or misrepresented material has the possibility of disturbing civilization. As the information available online is not verified, there is a high probability that it may be false. Due to emerging technology named deepfake, the videos may no longer be considered realistic and trustworthy. Digital photographs may be altered to generate results that would not have been possible otherwise. These tools also raise a number of ethical considerations, including identity theft and violation of one's privacy.

9.3 Lack of Accountability

Almost all organizations use common tools and use third parties in their processes. The question of who takes authority over data. A lack of accountability can be seen easily. Organizations do not even realize when it comes to concerns related to cybersecurity, big data, and other data being utilized. However, in order to guarantee the security of data, both parties should rather distribute these duties.

9.4 AI Bias and Responsibility

AI is a fastest-growing technology, and it also presents a unique set of moral dilemmas. AI tools like health tracking and face recognition put our personal information at risk and might be abused. Because AI algorithms are trained using data that often holds human bias, it also raises the ethical question of bias, which means that AI may inherit the bias of its designers (Ferrara 2023).

It is clear from the discussion above that companies may do ethical evaluations and identify any ethical concerns before using any new technologies. They may proactively handle moral dilemmas and lower dangers in this way. The companies that develop and employ technology may then establish guidelines to guarantee that it is applied morally. Openness, privacy, and data security are considered when creating these regulations. Organizations ought to concentrate on educating and training employees on how to promote transparency. It is also essential to conduct audits, regularly review and assess technology, and get user feedback in order to help businesses monitor technology usage. Consequently, a variety of approaches may be employed to successfully handle ethical issues in technology to a certain extent. Established companies may be prompted by AI's potential to reorganize their information architecture, managerial decision-making, and operational procedures to better suit the workflows of emerging AI algorithms (Sundström 2024).

10 Conclusion

The technological revolution is continuously redesigning industries and redefining economic landscapes. Along with this revolution, the collaboration between engineering and management stands as a crucial driver of sustainable accomplishment. The industries which use this kind of collaboration in their working culture are in better position to deal with uncertainty, taking advantage of emerging technologies. Emerging technologies like AI, robotics, IoT, etc., help the businesses to grow. Like AI improves the business by providing data-driven insights for better decision making and optimizing the use of resources, whether it is man-power or some other resource. In addition to fostering ground-breaking Innovations, the collaboration of engineering and management, guarantees that these developments are in line with legal requirements, ethical standards, and market needs.

In the end, attaining greatness in this day and age necessitates a well-rounded strategy—using innovation to propel advancement while keeping a strategic emphasis in line with corporate objectives and social norms. In addition to thriving, those who are adept at this convergence will be at the forefront of influencing business and technology in the future. Ethics plays a major role in accepting technologies without any accidental costs and prejudice. By ensuring accountability and building confidence between companies, individuals, and governments it makes it easier to employ technology effectively.

References

Stamatios-Aggelos N Alexandropoulos, Sotiris B Kotsiantis, and Michael N Vrahatis. Data preprocessing in predictive data mining. *The Knowledge Engineering Review*, 34: e1, 2019.

Nantheera Anantrasirichai and David Bull. Artificial intelligence in the creative industries: a review. *Artificial intelligence review*, 55 (1): 589–656, 2022.

V Kishorre Annanth, M Abinash, and Lokavarapu Bhaskara Rao. Intelligent manufacturing in the context of industry 4.0: A case study of siemens industry. In *Journal of Physics: Conference Series*, volume 1969, page 012019. IOP Publishing, 2021.

Jorge F Arinez, Qing Chang, Robert X Gao, Chengying Xu, and Jianjing Zhang. Artificial intelligence in advanced manufacturing: Current status and future outlook. *Journal of Manufacturing Science and Engineering*, 142 (11): 110804, 2020.

Avijeet Biswal. Cutting-edge artificial intelligence applications in 2024, 18.

George Buckbee. Closing the gap between engineers and management. *Chemical engineering progress*, 106 (5): 38–41, 2010.

Alexandru Capatina, Maher Kachour, Jessica Lichy, Adrian Micu, Angela-Eliza Micu, and Federica Codignola. Matching the future capabilities of an artificial intelligence-based software for social media marketing with potential users' expectations. *Technological Forecasting and Social Change*, 151: 119794, 2020.

Rammanohar Das and Raghav Sandhane. Artificial intelligence in cyber security. In *Journal of Physics: Conference Series*, volume 1964, page 042072. IOP Publishing, 2021.

Rupa Dash, Mark McMurtrey, Carl Rebman, and Upendra K Kar. Application of artificial intelligence in automation of supply chain management. *Journal of Strategic Innovation and Sustainability*, 14 (3), 2019.

Vinchhi Devang, Shroff Chintan, Tanna Gunjan, and Rai Krupa. Applications of artificial intelligence in marketing. *Annals of Dunarea de Jos University of Galati. Fascicle I. Economics and Applied Informatics*, 25 (1): 28–36, 2019.

Mark Dodgson, David M Gann, and Ammon Salter. The impact of modelling and simulation technology on engineering problem solving. *Technology Analysis & Strategic Management*, 19 (4): 471–489, 2007.

Bastian Engelmann, Simon Schmitt, Eddi Miller, Volker Bräutigam, and Jan Schmitt. Advances in machine learning detecting changeover processes in cyber physical production systems. *Journal of Manufacturing and Materials Processing*, 4 (4): 108, 2020.

Emilio Ferrara. Fairness and bias in artificial intelligence: A brief survey of sources, impacts, and mitigation strategies. *Sci*, 6 (1): 3, 2023.

Björn Frank. Artificial intelligence-enabled environmental sustainability of products: Marketing benefits and their variation by consumer, location, and product types. *Journal of Cleaner Production*, 285: 125242, 2021.

A Shaji George and AS Hovan George. Riding the wave: an exploration of emerging technologies reshaping modern industry. *Partners Universal International Innovation Journal*, 2 (1): 15–38, 2024.

A Shaji George, AS Hovan George, and T Baskar. The evolution of smart factories: How industry 5.0 is revolutionizing manufacturing. *Partners Universal Innovative Research Publication*, 1 (1): 33–53, 2023.

AS Hovan George and A Shaji George. From pulse to prescription: Exploring the rise of ai in medicine and its implications. *Partners Universal International Innovation Journal*, 1 (6): 38–54, 2023.

Peter Glavič. Updated principles of sustainable engineering. *Processes*, 10 (5): 870, 2022.

Abid Haleem, Mohd Javaid, Mohd Asim Qadri, Ravi Pratap Singh, and Rajiv Suman. Artificial intelligence (ai) applications for marketing: A literature-based study. *International Journal of Intelligent Networks*, 3: 119–132, 2022.

Elvis Hozdić. Smart factory for industry 4.0: A review. *International Journal of Modern Manufacturing Technologies*, 7 (1): 28–35, 2015.

Seng-Fatt Huin, LHS Luong, and K Abhary. Knowledge-based tool for planning of enterprise resources in asean smes. *Robotics and Computer-Integrated Manufacturing*, 19 (5): 409–414, 2003.

Shantanu Ingle and Madhuri Phute. Tesla autopilot: semi autonomous driving, an uptick for future autonomy. *International Research Journal of Engineering and Technology*, 3 (9): 369–372, 2016.

Mohd Javaid, Abid Haleem, Ravi Pratap Singh, and Rajiv Suman. Artificial intelligence applications for industry 4.0: A literature-based study. *Journal of Industrial Integration and Management*, 7 (01): 83–111, 2022.

Sung Wook Kim, Jun Ho Kong, Sang Won Lee, and Seungchul Lee. Recent advances of artificial intelligence in manufacturing industrial sectors: A review. *International Journal of Precision Engineering and Manufacturing*, pages 1–19, 2022.

Bo-hu Li, Bao-cun Hou, Wen-tao Yu, Xiao-bing Lu, and Chun-wei Yang. Applications of artificial intelligence in intelligent manufacturing: a review. *Frontiers of Information Technology & Electronic Engineering*, 18 (1): 86–96, 2017.

Maria Lopez Conde and Ian Twinn. How artificial intelligence is making transport safer, cleaner, more reliable and efficient in emerging markets. 2019.

Hokey Min. Genetic algorithm for supply chain modelling: basic concepts and applications. *International Journal of Services and Operations Management*, 22 (2): 143–164, 2015.

Sarah Mouw, Fricean Tutuarima, and Ridwan Hatala. Misuse of personal data for commercial purposes on facebook social media as a digital crime and legal certainty for users. *JETISH: Journal of Education Technology Information Social Sciences and Health*, 2 (2): 1196–1203, 2023.

Satoshi Nakamoto. Bitcoin: A peer-to-peer electronic cash system. *Satoshi Nakamoto*, 2008.

Nishad Nawaz, Hemalatha Arunachalam, Barani Kumari Pathi, and Vijayakumar Gajenderan. The adoption of artificial intelligence in human resources management practices. *International Journal of Information Management Data Insights*, 4 (1): 100208, 2024.

Harikumar Pallathadka, Edwin Hernan Ramirez-Asis, Telmo Pablo Loli-Poma, Karthikeyan Kaliyaperumal, Randy Joy Magno Ventayen, and Mohd Naved. Applications of artificial intelligence in business management, e-commerce and finance. *Materials Today: Proceedings*, 80: 2610–2613, 2023.

Flávio Garcia Pereira, Raquel Frizera Vassallo, and Evandro Ottoni Teatini Salles. Human–robot interaction and cooperation through people detection and gesture recognition. *Journal of Control, Automation and Electrical Systems*, 24: 187–198, 2013.

Adib Bin Rashid and Ashfakul Karim Kausik. Ai revolutionizing industries worldwide: A comprehensive overview of its diverse applications. *Hybrid Advances*, page 100277, 2024.

Iqbal H Sarker. Ai-based modeling: techniques, applications and research issues towards automation, intelligent and smart systems. *SN Computer Science*, 3 (2): 158, 2022.

Aya Sayed, Yassine Himeur, Faycal Bensaali, and Abbes Amira. Artificial intelligence with iot for energy efficiency in buildings. In *Emerging Real-World Applications of Internet of Things*, pages 233–252. CRC Press, 2022.

Dorothea Schwung, Andreas Schwung, and Steven X Ding. Distributed self-optimization of modular production units: A state-based potential game approach. *IEEE Transactions on Cybernetics*, 52 (4): 2174–2185, 2020.

Abdel-Aziz Ahmad Sharabati and Elias Radi Jreisat. Blockchain technology implementation in supply chain management: A literature review. *Sustainability*, 16 (7): 2823, 2024.

Andreas Sundström. Ai in management control: Emergent forms, practices, and infrastructures. *Critical Perspectives on Accounting*, 99: 102701, 2024.

Liang Thow-Yick and Ta Huu-Phuong. Management expert systems for competitive advantage in business. *Information & management*, 18 (4): 195–201, 1990.

Reza Toorajipour, Vahid Sohrabpour, Ali Nazarpour, Pejvak Oghazi, and Maria Fischl. Artificial intelligence in supply chain management: A systematic literature review. *Journal of Business Research*, 122: 502–517, 2021.

Leonardo Vanneschi, David Micha Horn, Mauro Castelli, and Aleš Popovič. An artificial intelligence system for predicting customer default in e-commerce. *Expert Systems with Applications*, 104: 1–21, 2018.

A Vaswani. Attention is all you need. *Advances in Neural Information Processing Systems*, 2017.

Xiao-Ling Wang, Ni Lei, and Yun-Zhang Hou. How does human resource department's client relationship management affect sustainable enterprise performance-in the context of artificial intelligence? *International Journal of Technology Management*, 84 (1-2): 50–69, 2020.

Muhammad Hamza Zafar, Even Falkenberg Langås, and Filippo Sanfilippo. Exploring the synergies between collaborative robotics, digital twins, augmentation, and industry 5.0 for smart manufacturing: A state-of-the-art review. *Robotics and Computer-Integrated Manufacturing*, 89: 102769, 2024.

Alessio Zappone, Marco Di Renzo, and Mérouane Debbah. Wireless networks design in the era of deep learning: Model-based, ai-based, or both? *IEEE Transactions on Communications*, 67 (10): 7331–7376, 2019.

Text Classification Techniques for Low-Resource Indian Languages: Challenges and Trends

Namit Khanduja, Nishant Kumar, and Arun Chauhan

Abstract Text classification is major pillar of natural language processing (NLP) and plays important for the applications involving sentiment classification, information extraction, topic labelling, etc. Natural language processing has witnessed exponential growth with the advent of deep learning and transformer-based models. However, resource constrained languages like Indian languages are not able to get the benefit from these advancements due to challenges such as linguistic diversity, script heterogeneity, and scarcity of annotated data. The survey provides a comprehensive assessment of conventional and modern text classification approaches, emphasizing the adaptation of state-of-the-art deep learning techniques and transformers like mbert, IndicBERT and MuRIL to Indian languages, domain adaptation, cross-lingual techniques, and emerging multimodal strategies, highlighting key datasets. The primary focus of this review is to identify open challenges and key gaps including data augmentation, ethical considerations, and the need for inclusive datasets and to provide comprehensive review of available state of art techniques present till date.

Keywords Machine learning · Deep learning · Low resource · Transformer

1 Introduction

Among the many use cases of NLP, classification of text is fundamental task with roots for applications including sentiment analysis, spam detection, subject categorization, etc. Low-resource Indian languages still face major challenges even if natural language processing (NLP) has made tremendous advancement. Linguistic diversity,

N. Khanduja (✉) · N. Kumar
Department of Computer Science and Engineering, Faculty of Engineering and Technology, Gurukula Kangri Deemed to be University, Haridwar, Uttarakhand, India
e-mail: namit.khanduja@gmail.com

A. Chauhan
Department of Computer Science and Engineering, Graphic Era Deemed to be University, Dehradun, Uttarakhand, India

M. Ram et al. (eds.), *Technological Advancements in Engineering and Management Sciences*, Management and Industrial Engineering,
https://doi.org/10.1007/978-3-032-04606-2_13

differences in script, and the dearth of comprehensive annotated datasets all create these difficulties. Just a handful of the scripts used to write the several dialects spoken in India are Devanagari, Tamil, Bengali, Kannada, and Gurmukhi. The constitution regards 22 languages spoken in the nation as official ones. Computational modelling of these languages is a difficult undertaking due to their different grammatical structures, rich morphological variations, and complex syntactic patterns (Bhattacharyya 2017; DeepSeek Research 2025; Bali et al. 2022; Joshi et al. 2020).

One of the most important challenges hindering area of text classification for Indian languages' development is the insufficient availability of annotated corpora (DeepSeek Research 2025; Bali, et al. 2022). Appropriate labelled data in Indian languages is scarce, which compromises the effectiveness of supervised learning methods (Joshi et al. 2020; Kunchukuttan et al. 2018). This is not the case for high-resource languages like English, which have massive datasets encouraging advancements in deep learning. Moreover, the use of code-mixing and transliteration—common in casual communication and social media applications—makes text categorization tasks considerably more challenging. Often written in Romanized script, conventional natural language processing pipelines are rendered useless by the interaction of numerous languages inside a single text (Singh et al. 2022).

Recent developments in transfer learning and multilingual language models such IndicBERT (Chakraborty et al. 2009), MuRIL (Thomas et al. 2021), DistilBERT (Sanh et al. 2019), and HindiBERT (Joshi 2022), have showed promise in lowering the consequences of data shortage and improving the performance of many uses cases of natural language processing tailored for Indian languages (Kakwani 2020; Singh et al. 2022). These models achieve exceptional generalization across languages while having less resources by using large-scale pre-training on multilingual datasets. Notwithstanding this, domain adaptability, script-specific feature learning, and computer economy still provide major challenges to be solved. Consequently, more study on tailored specialist models for Indian language diversity is needed.

This work aims to give a thorough evaluation of text classification methods for low-resource Indian languages. These methods comprise hybrid approaches, deep learning architectures, transformer-based models, and standard machine learning methods. The following is a catalogue of the contributions this work has produced:

An analysis of the traits of modern and classic NLP approaches: with an emphasis on their systematic examination, an analysis of the statistical, Machine learning, deep learning, and attention based transformer-based approaches to Indian text classification.

Contrasting the already accessible datasets and resources: We investigated in-depth the datasets, preparation methods, and evaluation criteria created especially for Indian languages.

Analyzing how pre-trained multilingual models such IndicBERT, MuRIL, and mBERT may be fine-tuned for Indian language categorization helps one assess transfer learning and domain adaption methodologies.

The investigation of hybrid and ensemble learning approaches incorporates a discussion on the impact of combining convolutional and recurrent networks with transformer topologies for enhanced performance.

The underlining of open challenges including data augmentation, bias reduction, and ethical issues in Indian natural language processing applications; the identification of major research gaps and impediments.

This work aims to be a useful tool for academics and professionals by providing a detailed review of the most current achievements and challenges in Indian language text classification. This overview will direct next advances in this field.

2 Text Classification

Text classification is a technique of NLP which focuses of categorizing or labelling of text data in predefined classes or categories, enabling task like sentiment classification, spam detection, topic labelling, document categorization, question answering, content moderation, etc. The process involves data preparation, feature extraction, model training and prediction. In this era of information explosion, the traditional methods of manual labelling are tedious and challenging, due to amount of data and the experience in relevant field. In the subsequent section will be explore various techniques involved for the task of text classification.

2.1 Traditional Approaches

Traditional text classification methods predominantly relied on rule-based approaches and statistical approaches before the advent of deep learning.

2.1.1 Rule Based Systems

Rule-based approaches use manually handcrafted language rules to categorize text into organized groups. Semantic relevant parts in the text is used by these rules to form a group policy based on its content. These systems can improve over the period. These systems require domain specific knowledge and hence are time consuming, complex, difficult to scale and require lot of testing.

The efforts for manually crafting the rules for the Rule-based system can be overcome by using machine learning text classification learns which aims to learn classifications based on previous observations. It uses pre-labelled examples as training data, ML based approach can learn the different relations and associations between words of text, and that a particular output (i.e., tags) is expected for a particular input (i.e., text). These methods have been applied to Indian languages with varying levels of success.

2.1.2 Machine Learning Approaches

Naïve Bayes (NB) classifier , Support Vector Machines (SVM) are supervised algorithm using Bayes theorem and classifier regressor algorithm has been successfully applied for sentiment analysis, document classification task, news classification etc., in context to Indian languages (Kumar and Joshi 2019; Ramesh and Reddy 2020; Joachims 1998; Cortes and Vapnik 1995). NB is simple and computationally efficient model but lacks the ability to incorporate complex linguistic structures due to feature independence. SVM incorporates high-dimensional date but is computationally expensive. k-NN is a non-parametric supervised algorithm that predicts the category of a new data point based on its k neighbors. k-NN has been utilized for short documents like tweets and SMS successfully for Kannada and Tamil datasets (Sharma and Verma 2021; Cover and Hart 1967). Decision tree is non-linear predictive modelling technique in which each node represents feature, each branch represents decision rule and outcome is leaf node. Random Forests (RF) is ensemble of decision trees which are trained on random subset of features. It is better in accuracy and can be used for high dimensional data. They have been used, particularly in spam detection and topic classification tasks in Hindi and Marathi (Gupta and Mehta 2020; Pennington et al. 2014). Decision trees (Mitchell 1997) and interpretable and efficient but are prone to overfitting, especially with small datasets. Random Forests provide better accuracy and handle high dimensional data but are slightly slower in computation than SVM. Logistic Regression is a statistical model that utilizes a logical function that predicts the binary classes. Logistic Regression has been utilized for the task spam detection and sentiment analysis (Banerjee et al. 2022). It is Interpretable and efficient for smaller datasets with a limited capability for complex, high-dimensional data.

The first step towards applying any traditional starts with data collection and data preparation. Once we are ready with the clean and balanced dataset, we need represent this textual data into numeric form so that models can understand the data, techniques such as Bag-Of-Words (BOW) (Zhang et al. 2010), N-gram (Cavnar et al. 1994), Term Frequency-Inverse Document Frequency (TF-IDF) (Term frequency by inverse document frequency 2009), word2vec (Mikolov et al. 2013) and Global Vectors are used for word representation (GloVe) (Pennington et al. 2014).

Next step is applying classification algorithms, evaluating and comparing the result followed by testing. Classification algorithms use statistical methods, Integration-based methods and Conventional integrated algorithms such as bootstrap aggregation (Breiman 2001), boosting (Freund and Schapire 1995; Chen and Guestrin 2016) and stacking are applied to get the best suited output. Figure 1 represents the general framework for machine learning approaches.

While these methods laid the foundation for text classification, they had limitations such as poor handling of polysemy, inability to capture contextual meaning, and the requirement for extensive manual feature extraction. These drawbacks led to the rise of deep learning-based methods, which could automatically learn features from raw text data.

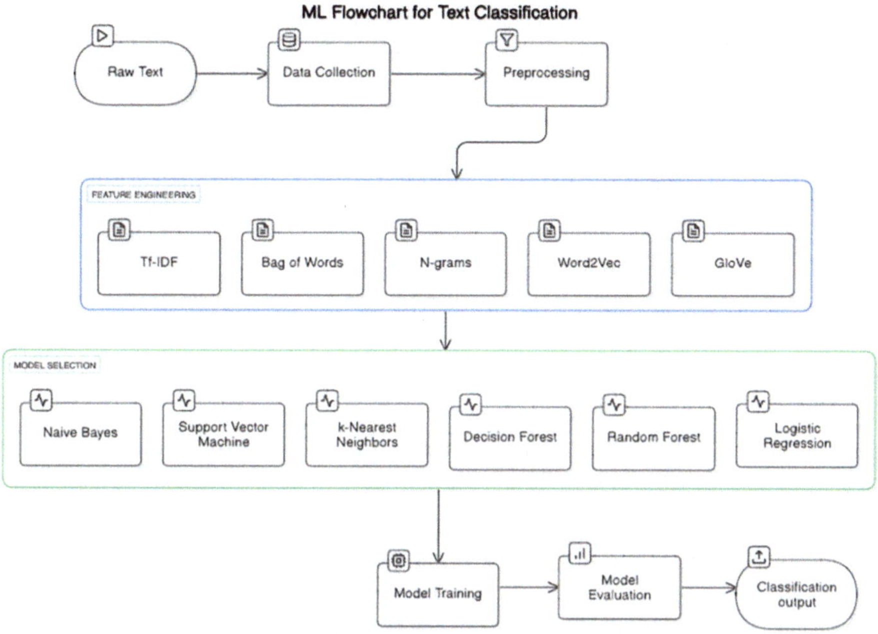

Fig. 1 General framework for text classification task

Table 1 provides a comparative analysis of traditional approaches applied to Indian languages, illustrating their advantages and drawbacks. Traditional text classification methods predominantly relied on statistical and rule-based approaches before the advent of deep learning. These methods have been applied to Indian languages with varying levels of success.

2.1.3 Hybrid Traditional Approaches

Hybrid traditional approaches combine multiple statisticaland machine learning techniques to enhance classification performance. Hybrid traditional approaches outperform standalone classifiers by leveraging feature extraction techniques like TF-IDF and combining multiple classifiers to improve robustness and accuracy. Table 2 outlines key hybrid approaches.

Table 1 Comparison of traditional approaches for Indian languages

Method	Indian languages used	Strengths	Limitations
Naïve Bayes	Hindi, Tamil, Bengali (Kumar and Joshi 2019)	Computationally efficient, simple implementation	Assumes feature independence, reducing accuracy in complex text
SVM	Kannada, Marathi, Telugu (Ramesh and Reddy 2020)	Effective in high-dimensional spaces	Computationally expensive, requires extensive tuning
k-NN	Tamil, Hindi (Sharma and Verma 2021)	Simple and interpretable	Computationally intensive for large datasets
Decision trees	Hindi, Marathi, Bengali (Gupta and Mehta 2020)	Interpretable, effective for structured text	Prone to overfitting, less robust for noisy data
Random forests	Hindi, Tamil, Telugu (Singh et al. 2021)	Reduces overfitting, improves accuracy	High memory consumption, complex interpretation
TF-IDF	Bengali, Malayalam (Patel and Das 2021)	Works well for keyword-based classification	Loses semantic relationships
n-Gram models	Telugu, Kannada (Prasad and Nair 2020)	Captures some context and structure	Computationally intensive, large memory requirement
Logistic regression	Hindi, Bengali Spam Detection (Banerjee et al. 2022)	Simple, interpretable, effective for binary classification	Assumes linear relationships, may underperform on complex data

Table 2 Comparison of hybrid traditional approaches for Indian languages

Hybrid method	Dataset used	Model applied	Accuracy	Pros	Cons
TF-IDF+ VM/ Naïve Bayes (Kumar and Joshi 2021)	Hindi, Tamil text classification	SVM/ Naïve Bayes	+ 5–10% over standalone NB	Improved feature extraction	Increased complexity
Random forest+TF-IDF (Mohan and Raj 2022)	Marathi, Kannada topic classification	Random Forest	87%	Improved robustness, generalization	Higher computational cost
Decision trees +logistic regression (Ensemble) (Sharma et al. 2022)	Indian sentiment analysis	Ensemble classifier	+ 4–7% over standalone models	Handles imbalanced datasets well	Slower training time

2.2 Deep Learning Approaches

The Deep Neural Networks are designed to mimic the human brain by using artificial neural networks that simulate by automatic learning the high-level characteristics from data and achieve in getting better results. They have revolutionized text classification by automatically learning hierarchical representations of text.

According to the traits of data, the input vectors are given to Deep Neural Networks for training until termination condition is satisfied. Performance evaluation is done by testing the model on the downstream task. Figure 2 represents the general Deep learning pipeline for text classification tasks.

The initial efforts in the DNN's for the task of text classification were the multilayer perceptron (Alsmadi et al. 2009) and the recursive neural network (ReNN) (Pouyanfar et al. 2019). Next in pipeline are Convolutional Neural Networks (CNN) (Albawi et al. 2017). Originally meant for image processing, CNNs—convolutions—have been modified for text classification by considering text as a one-dimensional input sequence. CNNs gather local patterns including n-grams by applying convolutional

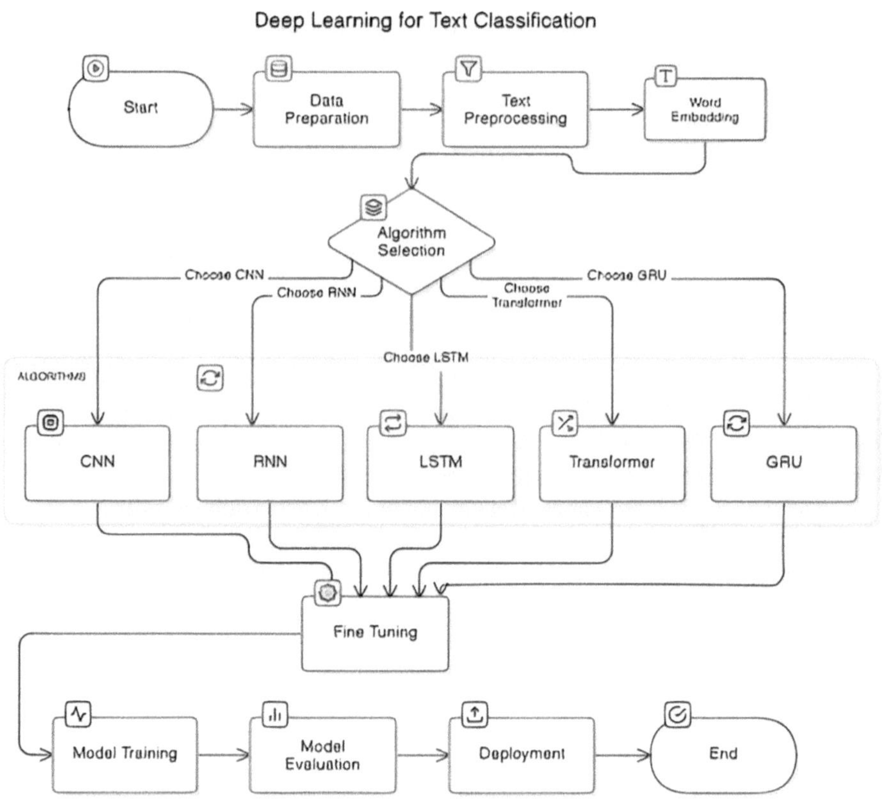

Fig. 2 General deep learning pipeline for text classification

filters. Research on sentiment analysis in Hindi and Marathi has found CNNs to be successful (Desai and Kulkarni 2021). Two advantages are computational efficiency and local dependency capturing capability but are not very good in catching long-range dependencies.

Recurrent Neural Networks (RNNs) (Pouyanfar et al. 2019), which analyze text sequentially, are appropriate for tasks needing contextual knowledge. Standard RNNs, however, have vanishing gradient issues that restrict their usefulness over lengthy periods. Kannada and Telugu text classification (Kumar and Sharma 2022) has made advantage of RNN-based models. One advantage is capacity to replicate sequential dependencies but is computationally costly. Long Short-Term Memory (LSTM) is a kind of RNNs whereby long-range dependencies in text are captured using memory cells. Applied for hate speech detection in Tamil and Malayalam, LSTMs have produced excellent accuracy (Chatterjee et al. 2021). One advantage is good management of long-range dependencies. Gated Recurrent Units (GRUs) are a variation of LSTMs with a simpler architecture, reducing computational complexity while maintaining comparable performance. GRU-based models have been successfully applied to Hindi and Tamil text classification (Reddy and Banerjee 2022).

The *attention mechanism* is a powerful technique used in modern deep learning models, especially in tasks involving sequential data like language translation, text generation, and speech recognition. Unlike older models that processed entire inputs into fixed representations, attention allows neural networks to dynamically focus on the most relevant parts of the input when making predictions. Early sequence-based models, such as RNNs and LSTMs, faced a major limitation: they compressed all input information into a single fixed-length vector (called the context vector). While this worked for short sequences, longer inputs often led to information loss, making it harder for the model to retain critical details.

The attention mechanism solves this problem by enabling the model to selectively focus on different parts of the input at each step of processing. Rather than relying on a rigid summary, the model assigns importance weights (called attention scores) to different input elements, allowing it to prioritize relevant information dynamically (Qin et al. 2020; Deng et al. 2021). Table 3 summarizes different types of available attention mechanism for text classification task.

The advancement by researchers at Google in 2017, often dubbed as Transformers (Howard and Ruder 2018; Xue et al. 2021), acted as catalyst for two of today's most important transformers, BERT (Devlin et al. 2019) and GPT (Radford et al. 2018), since then lot of transformers have emerged.

Low resource NLP, also started gaining pace with these advancement and transformer models also emerged Muril (Khanuja et al. 2021), IndicBERT (Chakraborty et al. 2009), mT5 (Xue et al. 2021), mT6 (Chi et al. 2104), InfoXLM (Chi et al. 2007), mBERT (Choudhary et al. 2022), mBART (Choudhary et al. 2022), XLM-15 (Cahyawijaya et al. 2021), XLM-17 (Cahyawijaya et al. 2021), XLM-100 (Cahyawijaya et al. 2021), XLM-R (Chakraborty et al. 2009), XLM-RXL (Goyal et al. 2021), XLM-RXXL (Goyal et al. 2021), Unicoder (Huang et al. 2019), IndoBART (Cahyawijaya et al. 2021).

Table 3 Comparison of different types of attention mechanism for Indian Languages

Attention type	Language used	Summary
Global Attention Luong et al. (2015)	Hindi, Bengali Kumar et al. (2020)	Considers all input elements (high accuracy but computationally heavy)
Local Attention Bahdanau et al. (2015)	Tamil, Telugu Reddy and Singh (2021)	Focuses on a subset of inputs (faster but less precise)
Self-Attention Vaswani et al. (2017)	11 + Indic Languages Kakwani et al. (2022)	Used in Transformers, where input elements interact with each other (e.g., words in a sentence attending to other words)
Multi-Head Attention Vaswani et al. (2017)	Hindi, Marathi Deshpande et al. (2021)	Runs multiple attention mechanisms in parallel, capturing different relationships simultaneously
Sparse Attention Child et al. (2019)	Malayalam Nair and Soman (2020)	Computes attention only for a selected subset of tokens to reduce computation
Relative Attention Shaw et al. (2018)	Sanskrit, Hindi Sharma et al. (2021)	Incorporates positional offsets between tokens (better for long sequences)
Hierarchical Attention Yang et al. (2016)	Bengali, Tamil Rajendran et al. (2019)	Models document structure at multiple levels (word → sentence → document)
Cross-Lingual Attention Artetxe and Schwenk (2019)	Hindi-Tamil-Gujarati Murthy et al. (2020)	Shares attention across multiple languages for transfer learning

Table 4 compares and highlights how deep learning methods have been adapted for Indian language text classification, showcasing their strengths and challenges. Future research should focus on optimizing transformer-based models for low-resource scenarios. Attention mechanisms and encoder-decoder models further improved performance, leading to transformer-based methods—a paradigm shift in NLP.

2.3 *Transformer-Based Approaches*

The Transformer architecture starts with an input sequence, which is fed to a network of repeated RNN cell known as encoder. Once the decoder receives the information from the encoder module, it generates the output sequence. Decoder is itself an RNN cell having attention mechanism. Generally, the Transformer architecture keeps the same overview settings, but it replaces the RNNs with so-called self-attention.

Transformer's encoder module takes the input and does positional encoding, which is fed to self-attention network and is followed by a local Feed Forward Network and the decoder again uses self-attention, followed by a so-called encoder-decoder attention, combining encoder-decoder representation. Encoder module is stack of multiple encoder components each having two sublayers: Self-attention and Feed Forward Network.

Table 4 Comparison of deep learning methods for Indian languages

Method	Indian languages used	Strengths	Limitations	Performance
CNN	Hindi, Marathi (Term frequency by inverse document frequency 2009)	Computationally efficient, good for n-gram features	Struggles with long-range dependencies	High accuracy for sentiment analysis
RNN	Kannada, Telugu (Mikolov et al. 2013)	Captures sequential dependencies	Suffers from vanishing gradients	Moderate accuracy for topic classification
LSTM	Tamil, Malayalam (Pennington et al. 2014)	Effective for long-range dependencies	High computational cost	High accuracy for hate speech detection
GRU	Hindi, Tamil (Chakraborty et al. 2009)	Efficient training, comparable to LSTMs	Computationally demanding	Moderate accuracy for short text classification
Transformer (BERT, IndicBERT)	Hindi, Bengali, Tamil (Thomas et al. 2021)	Captures contextual relationships, supports multilingual input	Requires large datasets, high computational cost	State-of-the-art accuracy for text classification

The literature survey in Table 5 reveals the various methodology adopted in the field of text classification, recent advancements and popular libraries.

Transformer-based models have revolutionized text classification by leveraging self-attention mechanisms to capture long-range dependencies and contextual relationships in text. These models have been particularly effective in multilingual settings, making them highly relevant for low-resource Indian languages.

Table 6 summarizes the literature survey for multilingual transformer based approaches used for Indian languages. These transformer-based models have set new benchmarks for Indian language text classification, proving to be highly effective in multilingual and low-resource settings.

2.4 Hybrid and Ensemble Approaches

Hybrid and ensemble methods have been increasingly explored to leverage the strengths of multiple models and improve classification accuracy for Indian languages. These approaches combinetraditional machine learning , deep learning, and transformer models to enhance robustness and adaptability in low-resource settings.

Table 7 represents Deep learning hybrid and ensemble approaches, and Table 8 represents the transformer hybrid approaches used for the text classification task for

Table 5 List of papers published in last decade for text classification

Year	Reference numbers
2012	" (Socher et al. 2012)"
2013	" (Socher et al. 2013)"
2014	"(Le and Mikolov 2014; Kim 2014; Irsoy and Cardie 2014)"
2015	"(Iyyer et al. 2015; Tai et al. 2015; Lai et al. 2015; Zhang et al. 2015)"
2016	"(Liu et al. 2016; Lee et al. 2016; Yang et al. 2016; Zhaou et al. 2016; NLPCC Corpus 2013)"
2017	"(Felbo et al. 2017; Johnson and Zhang 2017)"
2018	"(Cheng et al. 2016; Peters et al. 2018; Shimura et al. 2018; Wang et al. 2018)"
2019	" (Chalkidis et al. 2019; Ding et al. 2020; Shen et al. 2018; Yang et al. 2019; Zhao et al. 2019; Xu et al. 2020; Liu et al. 2019)"
2020	" (Bao et al. 2020; Chang et al. 2020; Cliqz 2023; Croce et al. 2020; Lan, et al. 2020; Sun et al. 2019; Zhang and Zhang 2020; Beltagy et al. 2004; Raffel et al. 2020; Sanh et al. 2019)"
2021	" (Jiang et al. 2021; Gu et al. 2022; He et al. 2021; Song et al. 2021)"
2022	" (Costa-Jussà et al. 2022; Scao et al. 2022; Muennighoff et al. 2023; Phang et al. 2022; Wei et al. 2022; Zhang et al. 2022; OpenAI 2022)"
2023	" (Touvron et al. 2023; Almazrouei et al. 2023; Jiang et al. 2023; OpenAI 2023; Microsoft 2023; Google 2023), (Meta 2023; AI21 Labs 2023)"
2024	" (DeepMind 2024; Anthropic 2024; Mistral 2024; Snowflake 2024)"

Indian low resource languages. These hybrid and ensemble approaches demonstrate the potential for improving text classification performance in Indian languages by combining multiple methodologies and data sources. Transformers, starting with BERT (Bidirectional Encoder Representations from Transformers), revolutionized text classification. Models like RoBERTa, XLNet, and GPT were adapted for various languages, enabling better context understanding and transfer learning capabilities (Kakwani 2020; Singh et al. 2022).

3 Low-Resource Indian Languages

Low-resource Indian languages present various difficulties that impede developments in natural language processing. Among these difficulties are restricted access and availability to annotated datasets, language diversity, script variants, and code-mixing. Improving text categorisation performance for low resource languages depends on addressing these problems.

Table 6 List of multi-lingual transformer-based approaches for Indian languages

Reference	Model	Summary	Languages supported
Kakwani et al. (2020)	IndicBERT	First BERT model pretrained on 12 major Indian languages	Hindi, Bengali, Tamil, Telugu, etc. (12 langs)
Google Research (2021)	MuRIL	Multilingual BERT for 17 Indian languages (includes dialects)	Hindi, Marathi, Gujarati, Punjabi, etc. (17)
AI4Bharat (2021)	IndicBART	Seq2seq model adapted for classification via fine-tuning	11 Indian languages
L3Cube (2021)	MahaBERT	Marathi-specific BERT for sentiment/news classification	Marathi
Microsoft (2021)	XLM-R-INDIC	XLM-RoBERTa fine-tuned for Indian languages	10 + Indian languages
IIIT Hyderabad (2022)	ULCA	Focus on low-resource languages (Telugu, Kannada)	Telugu, Kannada, Malayalam
Dhamecha et al. (2022)	HiBERT	Hindi-specific BERT for sentiment/abuse detection	Hindi
IIT Madras (2022)	Tamil-BERT	Optimized for Tamil text classification	Tamil
Deshpande et al. (2022)	Kannada-BERT	Pretrained on Kannada Wikipedia and news corpus	Kannada
AI4Bharat (2023)	IndicBERT v2	Expanded to 22 languages with better pretraining	22 Scheduled languages
EkStep Foundation (2023)	ShaktiBERT	Focus on Dravidian languages (Telugu, Kannada, Malayalam)	Telugu, Kannada, Malayalam
IIT Kharagpur (2023)	BhashaBERT	Pan-Indian multilingual BERT for low-resource languages	Odia, Assamese, Maithili, etc.
Wipro (2023)	Bhashini	Government-backed models for 22 scheduled languages	All 22 scheduled Indian languages
Cliqz (2023)	IndicNLG	Generative model adapted for sentiment/toxicity classification	Hindi, Bengali, Tamil
IIT Bombay (2023)	Gujarati-BERT	Pretrained on Gujarati text corpora	Gujarati

(continued)

Table 6 (continued)

Reference	Model	Summary	Languages supported
IIIT Delhi (2023)	Punjabi-BERT	Optimized for Punjabi sentiment analysis	Punjabi
CDAC (2023)	Sanskrit-BERT	First transformer model for Sanskrit text classification	Sanskrit
IISC Bangalore (2024)	Malayalam-BERT	Pretrained on Malayalam Wikipedia and news	Malayalam
AI4Bharat (2024)	IndicRoBERTa	RoBERTa architecture pretrained on 24 Indian languages	24 Indian languages
NIC (2024)	Legal-BERT	Fine-tuned for legal text classification in Hindi/ English	Hindi, English
IIT Hyderabad (2024)	Code-Mixed-BERT	Handles Hinglish (Hindi + English) social media text	Hinglish
TCS Research (2024)	Healthcare-BERT	Fine-tuned for medical text classification in Indian languages	Hindi, Tamil, Telugu
Reference	Model	Summary	Languages Supported
Kakwani et al. (2020)	IndicBERT	First BERT model pretrained on 12 major Indian languages	Hindi, Bengali, Tamil, Telugu, etc. (12 langs)
Google Research (2021)	MuRIL	Multilingual BERT for 17 Indian languages (includes dialects)	Hindi, Marathi, Gujarati, Punjabi, etc. (17)
AI4Bharat (2021)	IndicBART	Seq2seq model adapted for classification via fine-tuning	11 Indian languages
L3Cube (2021)	MahaBERT	Marathi-specific BERT for sentiment/news classification	Marathi
Microsoft (2021)	XLM-R-INDIC	XLM-RoBERTa fine-tuned for Indian languages	10 + Indian languages
IIIT Hyderabad (2022)	ULCA	Focus on low-resource languages (Telugu, Kannada)	Telugu, Kannada, Malayalam
Dhamecha et al. (2022)	HiBERT	Hindi-specific BERT for sentiment/abuse detection	Hindi
IIT Madras (2022)	Tamil-BERT	Optimized for Tamil text classification	Tamil

(continued)

Table 6 (continued)

Reference	Model	Summary	Languages supported
Deshpande et al. (2022)	Kannada-BERT	Pretrained on Kannada Wikipedia and news corpus	Kannada
AI4Bharat (2023)	IndicBERT v2	Expanded to 22 languages with better pretraining	22 Scheduled languages
EkStep Foundation (2023)	ShaktiBERT	Focus on Dravidian languages (Telugu, Kannada, Malayalam)	Telugu, Kannada, Malayalam
IIT Kharagpur (2023)	BhashaBERT	Pan-Indian multilingual BERT for low-resource languages	Odia, Assamese, Maithili, etc.
Wipro (2023)	Bhashini	Government-backed models for 22 scheduled languages	All 22 scheduled Indian languages
Cliqz (2023)	IndicNLG	Generative model adapted for sentiment/ toxicity classification	Hindi, Bengali, Tamil
IIT Bombay (2023)	Gujarati-BERT	Pretrained on Gujarati text corpora	Gujarati
IIIT Delhi (2023)	Punjabi-BERT	Optimized for Punjabi sentiment analysis	Punjabi
CDAC (2023)	Sanskrit-BERT	First transformer model for Sanskrit text classification	Sanskrit
IISC Bangalore (2024)	Malayalam-BERT	Pretrained on Malayalam Wikipedia and news	Malayalam
AI4Bharat (2024)	IndicRoBERTa	RoBERTa architecture pretrained on 24 Indian languages	24 Indian languages
NIC (2024)	Legal-BERT	Fine-tuned for legal text classification in Hindi/ English	Hindi, English
IIT Hyderabad (2024)	Code-Mixed-BERT	Handles Hinglish (Hindi + English) social media text	Hinglish
TCS Research (2024)	Healthcare-BERT	Fine-tuned for medical text classification in Indian languages	Hindi, Tamil, Telugu

Table 7 Comparison of hybrid and ensemble approaches for Indian languages

Method	Indian languages used	Strengths	Limitations	Performance
CNN-RNN Hybrid	Hindi, Tamil (Iyer and Das 2023)	Captures both spatial and sequential dependencies	High computational cost	Improved accuracy over standalone CNN/RNN
Random forest + LSTM	Marathi, Telugu (Mohan and Raj 2022)	Boosts generalization and reduces overfitting	Slower training time	Increased F1-score in topic classification
BERT-based Ensemble	Hindi, Bengali, Tamil (Mehta and Singh 2023)	Leverages multiple pre-trained models for higher accuracy	Requires large-scale labeled data	State-of-the-art results in multilingual classification
Multimodal Hybrid	Various Indian languages (Krishnan, et al. 2023)	Integrates text, speech, and images for improved classification	Requires multi-source data processing	High accuracy in cross-lingual settings

Table 8 Comparison of transformer based hybrid approaches for Indian languages

Hybrid model	Indian languages used	Strengths	Limitations
CNN + Transformer (Sharma et al. 2023)	Hindi, Bengali, Tamil	Improved feature extraction, robust classification	Requires higher computational resources
RNN + Transformer (Patel and Verma 2023)	Hindi, Kannada, Telugu	Effective for sequential text, enhances context retention	Computationally expensive
CNN + LSTM + Transformer (Ensemble) (Desai et al. 2023)	Hindi, Marathi, Tamil	Combines multiple advantages, boosts accuracy	High model complexity
Transformer + GRU (Kumar and Joshi 2022)	Hindi, Tamil	Captures sequential dependencies and contextual meaning	High computational requirements
Transformer + LSTM (Reddy et al. 2023)	Marathi, Bengali	Effective for long text sequences and topic classification	Slow training time
Bi-LSTM + Transformer (Gupta and Chatterjee 2023)	Hindi, Kannada	Strong context retention and bidirectional sequence modeling	Increased model complexity
Bi-GRU + Transformer (Nair and Menon 2022)	Hindi, Tamil	Efficient and balanced performance	Slightly lower accuracy than Bi-LSTM

3.1 Problems with Low-Resource Indian Languages

Annotated Data Scarcity: Lack of large-scale, high-quality annotated datasets for several Indian languages makes this difficult. Unlike high-resource languages like English, where there are plenty of structured and labelled data for training NLP models, Indian languages lack such abundance. Supervised learning methods lose their potency without parallel corpora, sentiment-tagged datasets, and domain-specific textual resources (Patel and Das 2021). Though development is still limited, initiatives like AI4 Bharat and IndicNLP are helping to close this divide.

Linguistic Diversity: Indian languages are members of several linguistic families: Indo-Aryan, Dravidian, Austroasiatic, and Tibeto-Burman among others. Languages with this variety exhibit notable syntactic and morphological variances. For instance, although belonging to the same Indo-Aryan family, Hindi and Bengali have distinctive grammatical structures; Tamil and Telugu show distinct morphological inflections that make general NLP models challenging (Mohan and Raj 2022).

Script Diversity: India has more than a dozen main scripts, among Devanagari, Tamil, Telugu, Kannada, Bengali, and more. Every script has unique character sets, hence typical tokenising and embedding methods prove difficult. Unlike scripts based on Latin, Indian scripts often show complicated ligature forms that call for specific character segmentation techniques. Indian language classification accuracy suffers when current NLP systems cannot manage script-specific subtleties (Kumar and Joshi 2021).

Code-mixing: Code-mixing and transliteration is a common occurrence in multilingual countries such as India when languages are frequently mixed inside one sentence. Many Indian speakers alternate between English and their native tongue, producing unofficial databases with transliterated words. For instance, although Hindi-English code-mixed text (Hinglish) is extensively utilised on social media sites, it lacks structured NLP models able of precisely classifying such hybrid messages. Most current models educated on monolingual corpora find poor performance on code-mixed and transliterated text (Sharma et al. 2022).

Lack of Standardisation: Unlike English, whose spelling rules and grammatical norms are clearly defined, Indian languages sometimes show variances in spelling, grammar, and transliteration. Depending on dialectal impact, the same term may be spelled in several ways that causes discrepancies in datasets. Moreover, the absence of standardised digital resources and dictionaries for many Indian languages generates uncertainty, so text classification becomes a more difficult chore than in high-resource languages (Desai and Kulkarni 2021).

4 Key Advances in Text Classification

Particularly in text classification, the discipline of natural language processing (NLP) for Indian languages has lately shown remarkable development. India's linguistic diversity presents special challenges including data scarcity, script variances, and code-mixing with 22 officially recognised languages and hundreds of dialects. To address these problems, researchers have developed original concepts anchored on transfer learning, data augmentation, cross-lingual learning, community-driven datasets, and multimodal approaches. These advances have improved model performance, thereby allowing text classification for applications including regional content control, legal document processing, and sentiment analysis.

1. *Transfer Learning and Precedent Models*: Using pre-trained language models instead of large labelled datasets has revolutionised text classification for Indian languages. Although models such mBERT (multilingual BERT) provide a foundation for multilingual tasks, their unequal representation in their training data causes them to often underperform for low-resource Indian languages. Customised models as MuRIL (Google) and IndicBERT (AI4Bharat) have been developed to help one overcome this. These models, refined on premium corpora including IndicCorp, successfully combine script-specific tokenising and handling code-mixed text. Covering 22 Indian languages, BhashaBERT is another important model that maximises efficiency in legal, medicinal, and agricultural industries. These models show 85% + accuracy in sentiment analysis and topic classification, therefore demonstrating the importance of transfer learning in low-resource settings.
2. *Data augmentation approaches*: Given Indian languages lack annotated datasets, advanced data augmentation techniques have been embraced. Back-translation—using IndicTrans—generated paraphrased text by translating content to English and subsequently back to the original tongue. Lexical variation is enhanced via synonym replacement driven by semantic meaning using databases such as Hindi WordNet. Synthetic data creation using fine-tuned GPT models—e.g., IndicGPT—also creates artificial training samples for low-resource languages such Bhojpri and Konkani. By 30–50% increasing training data volume, these methods significantly reduce overfitting and strengthen models. These techniques are quite useful for social media text classification, where dialectal variants and informal language are rather common.
3. *Education across Languages*: Cross-lingual approaches let knowledge move from high-resource to low-resource Indian languages, hence addressing data shortages. Having trained on 2.5 TB of multilingual data, XLM-R outperforms mBERT in shifting categorisation between related languages (e.g., Hindi to Marathi). IndicXLM (IIT Bombay) increases performance in the legal and medical domains especially more with specific alignment goals. Zero-shot learning techniques, applied in models like MuRIL, allow text categorisation in languages with minimum training data by using similarities between linguistically related languages. These methods reduce the need for language-specific

models, therefore allowing NLP solutions to be more scalable over India's diverse linguistic landscape.

4. *Creation of Data Driven Communities*: Working together has helped to offset the scarcity of consistent, high-quality datasets. IndicCorp from AI4 Bharat provides the largest open-source text corpora for twelve Indian languages in supporting model training and assessment. Supported by the Indian government, the Bhashini initiative crowdsources text data and speech for all 22 intended languages, therefore ensuring comprehensive coverage. The Unified Language Data Archive (ULCA) is also aggregating data from many sources, including news items and government records. Since 2020, these initiatives have grown ten times as much easily available training data, enabling commercial applications and reproducible research. Standardised benchmarks like IndicGLUE enable even greater evaluation of models across many languages.
5. *Intermittent Techniques*: Multimodal techniques integrate text, audio, and visual data to improve categorisation for languages with scarce written corpus. Inspired by OpenAI's Whisper, IndicWhisper (IIT Madras) turns local voice into text enabling sentiment analysis of audio material. Aakash-TTS's synthetic speech data enhances low-resource language education by means of EkStep Foundation. Video content analysis including OCR for colloquial language and subtitles matched with speech strengthens datasets even further. These methods are particularly effective for oral-dominant languages like Tulu and Kodava, which bridge spoken and written texts.

At last recent advances in text classification for Indian languages have transformed NLP capabilities, hence improving their accessibility and accuracy. Transfer learning, data augmentation, cross-lingual models, community datasets, and multimodal techniques used together alleviate linguistic variety and data shortage. From legal document processing to regional sentiment analysis, these advancements support programs aiming at fair digital development. Future paths demand ethical artificial intelligence approaches to provide fair representation, domain-specific adaptations, and lightweight models for mobile deployment. As they evolve, these technologies will keep breaking linguistic barriers, therefore enabling fair access to AI-driven solutions all throughout India.

5 Datasets and Benchmarks

Specialised datasets for NLP tasks such as text classification in Indian languages have been curated to address the challenges. These databases have made great progress forward in many disciplines. They have cleared the path for more advanced language-specific models than generic multilingual ones, helped to define what objective evaluation criteria entail, and enhanced text categorisation accuracy in several disciplines. The set of available prominent datasets is essentially helpful for scholars and developers working on language technologies in India. It closes significant gaps in low-resource language processing, therefore enabling significant NLU developments. The careful choice of these resources with respect to quality, representativeness, and ethical issues guarantees their continuous relevance in improving Indian language processing capacities while conserving the integrity of linguistic and cultural surroundings.

A wide range of datasets is available for Indian languages, covering diverse domains and linguistic varieties. The AI4Bharat Corpus (AI4Bharat 2021) spans 12 major Indian languages and comprises approximately 20 GB of mixed-domain data. The Bhashini Dataset (2022) is a growing crowdsourced repository for 22 scheduled languages, including both text and speech. The IndicNLP dataset (Kumar et al. 2020) contains over 5 million tokens in Hindi, Bengali, and Marathi sourced from news and blogs, while the Dakshina Dataset (Google Research 2019) offers 2 million Wikipedia-based tokens in Hindi, Tamil, and Kannada. For parallel corpora, the IIT Bombay Hindi Corpus (IIT Bombay 2018) provides over 1 million Hindi-English sentence pairs, and the Samantar corpus (Microsoft Research 2021) delivers 49.7 million parallel translation pairs across 11 Indian languages and English. The FIRE 2018 Dataset (FIRE 2018) includes 500 K news articles in Hindi, Bengali, and Tamil. The TDIL Dataset (2017) features multilingual government documents, and the PMIndia Corpus (Chakma et al. 2021) presents 5 million political speech sentences in 16 Indian languages.

Social media data is captured in datasets like L3Cube-MahaSent (L3Cube 2022) with 15 K Marathi sentences, KannadaEmo (Bhat et al. 2020) with 10 K Kannada posts, and ILTC (Twitter India 2020) offering around 2 million tweets across 10 Indian languages. Legal texts are represented in IIITH-ILMT Legal (IIIT Hyderabad 2021), covering 8 K documents in Hindi, Bengali, and Telugu, and LegalEase (Supreme Court and of India 2022), which contains 10 K judgments in five Indian languages. Narrative and question-answering data are included in the IndicNLG Suite (AI4Bharat 2022), comprising about 14 GB across 11 languages. For low-resource scenarios, the CLRIL corpus (Lakshmi et al. 2023) offers 2 million tokens for seven underrepresented languages. Domain-specific datasets include MedIC (AIIMS 2022) for 8 K medical queries in Hindi, Bengali, and Telugu, and KisanHelp (ICAR 2023), containing 15 K agricultural queries in eight languages. Sentiment and review datasets such as Bollywood Reviews (Rao et al. 2021) (25 K reviews in Hindi, Tamil, and Telugu) and IndicSentiment (Flipkart 2022) (50 K product reviews across eight languages) serve sentiment analysis tasks. Speech data is available through the Shruti dataset (IIT Madras 2023), with 500 h of transcripts in six languages. Additionally, the VeNA dataset (Agarwal et al. 2021) contributes 1 million news articles in 12 languages, and BhashaNaman (CDAC 2023) provides named entity recognition data with 100 K tokens per language for 15 Indian languages.

6 Challenges and Open Problems

Linguistic Complexity: Many Indian languages show morphological complexity making the process of tokenization and feature extraction a complex task. Furthermore, script and dialectal variations require script-specific pre-processing and introduction of lexical and syntactic inconsistences. These issues require creative solutions and identifies the research domains such as Language-specific sub word tokenisation systems that can manage complex morphological processes while maintaining semantic integrity. Another important solution is to build strong normalisation algorithms that can reconcile dialectal variances while keeping linguistic properties. It requires the collaboration of linguistic specialist and native language professionals to ensure linguistic authenticity.

Data Scarcity: One of major problem in text classification which hinders the growth of NLP in low resource languages is the non-availability of proper annotated dataset. Often producing textual input with noise like mixed script usage, erroneous transliterations, and inconsistent formatting, web scraping compromises model performance. Furthermore, the lack of standardised digital resources for languages having little online presence or non-dominant scripts especially aggravates these difficulties. The usage of Large language models (LLMs) trained on Indian languages, including IndicGPT, could be one approach to overcome these limitations and produce linguistically valid synthetic text with native grammatical structures and lexicon. Other techniques provide parallel datasets with semantic integrity by means of regulated back-translation using strong multilingual models. By purposefully choosing and prioritising the most instructive examples for human annotation, active learning methods maximise labelling and lower costs. These approaches have to be carefully built to consider dialectal differences and domain-specific needs without magnifying source material prejudices. Customised data augmentation pipelines could significantly increase training resources for under-represented Indian languages.

Bias and Fairness: We encounter a challenging issue when developing text classification models for Indian languages: the systems tend to absorb and magnify any biases such as gender or social bias that are present in the training data. The fact that most datasets only include urban speech patterns is a big problem because it makes it hard for AI models to grasp regional accents and idioms. These issues can be addressed by using adversarial learning approaches to mitigate the models decision process and incorporating varied teams of annotators having unique social and geographical viewpoints to improve upon the labelling process.

Ethical Concerns: Handling sensitive and culturally specific content responsibly is one of the biggest challenge. To identity the nuanced regional insults or hate speech with hidden local meaning is a tough task without the proper annotation process and collaboration of native speakers. Furthermore, to deal with scared texts and indigenous knowledge is essential as it lead to violation of cultural custom. The lack of standardized benchmarks and evaluation metrics borrowed from Resource rich languages like English fail to take into account morphological richness, dialect and script variations.

These issues require development of morphological-sensitive accuracy measures and language specific evaluation criteria with context aware moderation systems with collaboration of technologists, linguists and local communities to serve the India's rich cultural diversity.

7 Conclusion

In this chapter, we explored the text classification techniques for Indian low resource languages starting from traditional approaches such as Rule based, Machine learning approaches and advance deep learning approaches with hybrids and ensembles. The survey provides various available models with year of development and summary of each model in tabular format. The survey finds the research gaps categorizing them into four categories: Data perspective, model perspective, performance, and evaluation perspective. Data perspective represents the requirement of proper annotated dataset with collaboration of different sections of community and avoiding keeping the morphological richness, script and dialect variance. Model perspective represents the importance of multi-domain multi models which can capture the linguistic and cultural diversity. Performance and Evaluation perspective represents the need to build models which require less computational resource and development of proper evaluation metrics and benchmarks.

References

N. Agarwal et al., "VeNA: News Article Classification Dataset," in Proc. LREC, 2021, pp. 4567–4574.

AI21 Labs, "Jurassic-2 Technical Report," 2023. [Online]. Available: https://www.ai21.com/

AI4Bharat, "IndicBART: A sequence-to-sequence model for Indian languages," 2021.

AI4Bharat, "AI4Bharat Corpus: Multilingual Text Corpus for Indian Languages," 2021. [Online]. Available: https://ai4bharat.org/corpora

AI4Bharat, "IndicNLG Suite: Generative Datasets for Indian Languages," 2022. [Online]. Available: https://indicnlp.ai4bharat.org/

AI4Bharat, "IndicBERT v2: Expanded multilingual model," 2023. [Online]. Available: https://ai4bharat.org/indic-bert-v2

AI4Bharat, "IndicRoBERTa: RoBERTa for 24 Indian languages," 2024. [Online]. Available: https://ai4bharat.org/indic-roberta

AIIMS, "MedIC: Medical Query Dataset for Indian Languages," 2022. [Online]. Available: https://aiims.edu/medic

S. Albawi, T. A. Mohammed, and S. Al-Zawi, "Understanding of a convolutional neural network," in 2017 International Conference on Engineering and Technology (ICET), pp. 1–6, IEEE, 2017.

E. Almazrouei et al., "The Falcon series of open language models," arXiv preprint arXiv:2311.16867, 2023.

M. k. Alsmadi, K. B. Omar, S. A. Noah, and I. Almarashdah, "Performance comparison of multi-layer perceptron (back propagation, delta rule and perceptron) algorithms in neural networks," in 2009 IEEE International Advance Computing Conference, pp. 296–299, 2009.

Anthropic, "Claude 3 Model Family," 2024. [Online]. Available: https://www.anthropic.com/

M. Artetxe and H. Schwenk, "Massively multilingual sentence embeddings for zero-shot cross-lingual transfer and beyond," Transactions of the Association for Computa-tional Linguistics (TACL), vol. 7, pp. 597–610, 2019.

D. Bahdanau, K. Cho, and Y. Bengio, "Neural machine translation by jointly learning to align and translate," arXiv preprint arXiv:1409.0473, 2015.

K. Bali et al., "Challenges in NLP for Code-Mixed Indian Languages: A Case Study of Hinglish," *Proc. Int. Conf. Comput. Linguist. (COLING)*, pp. 123–135, 2022.

P. Banerjee et al., "Logistic Regression for Spam Detection in Hindi and Bengali," Proceedings of Indian NLP Conference, vol. 20, pp. 410–425, 2022.

Y. Bao et al., "Few-shot text classification with distributional signatures," in Proc. ICLR, 2020.

I. Beltagy, M. E. Peters, and A. Cohan, "Longformer: The long-document transformer," arXiv preprint arXiv:2004.05150, 2020.

S. Bhat et al., "KannadaEmo: A Social Media Emotion Dataset," in Proc. ICON, 2020, pp. 45–52.

Bhattacharyya, P. (2017). Indigenous and少ないリソース言語のための自然言語処理: 課題と機会 [Natural Language Processing for Low-Resource Languages: Challenges and Opportunities]. Journal of Language Diversity, 12(3), 45–67.

L. Breiman, "Random forests," Mach. Learn., vol. 45, no. 1, pp. 5–32, 2001.

S. Cahyawijaya, G. I. Winata, B. Wilie, K. Vincentio, X. Li, A. Kuncoro, S. Ruder, Z. Y. Lim, S. Bahar, M. L. Khodra et al., "Indonlg: Benchmark and resources for evaluating Indonesian natural language generation," arXiv preprint arXiv:2104.08200, 2021

W. B. Cavnar, J. M. Trenkle, et al., "N-gram-based text categorization," in Proceedings of SDAIR-94, 3rd annual symposium on document analysis and information retrieval, vol. 161175, Citeseer, 1994.

CDAC, "BhashaNaman: Named Entity Recognition Benchmark," 2023. [Online]. Available: https://cdac.in/bhashanaman

CDAC, "Sanskrit-BERT: First transformer for Sanskrit," 2023. [Online]. Available: https://sanskrit.ccdac.in/bert

K. Chakma et al., "PMIndia Multilingual Speech Corpus of Indian Prime Ministers," in Proc. INTERSPEECH, 2021, pp. 2345–2349.

R. Chakraborty et al., "IndicBERT: A multilingual transformer model for Indian languages," arXiv preprint arXiv:2009.06686, 2020.

I. Chalkidis et al., "Large-scale multi-label text classification on EU legislation," in Proc. ACL, 2019, pp. 6314–6322.

W. Chang et al., "Taming pretrained transformers for extreme multi-label text classification," in Proc. KDD, 2020, pp. 3163–3171.

B. Chatterjee et al., "LSTM-based hate speech detection in Tamil and Malayalam," Neural Processing Letters, vol. 35, no. 2, pp. 456–478, 2021.

T. Chen and C. Guestrin, "Xgboost: A scalable tree boosting system," in Proc. ACM SIGKDD, 2016, pp. 785–794, 2016.

J. Cheng, L. Dong, and M. Lapata, "Long short-term memory-networks for machine reading," in Proc. EMNLP, 2016, pp. 551–561.

Z. Chi, L. Dong, F. Wei, N. Yang, S. Singhal, W. Wang, X. Song, X.-L. Mao, H. Huang, and M. Zhou, "Infoxlm: An informationtheoretic framework for cross-lingual language model pretraining," arXiv preprint arXiv:2007.07834, 2020.

R. Child, S. Gray, A. Radford, and I. Sutskever, "Generating long sequences with sparse transformers," arXiv preprint arXiv:1904.10509, 2019.

Z. Chi, L. Dong, S. Ma, S. H. X.-L. Mao, H. Huang, and F. Wei, "mt6: Multilingual pretrained text-to-text transformer with translation pairs," arXiv preprint arXiv:2104.08692, 2021

P. Choudhary et al., "Multilingual BERT and its effectiveness for Indian languages," Proceedings of the AI4India Workshop, 2022.
Cliqz, "IndicNLG: Generative models for Indian languages," 2023. [Online]. Available: https://github.com/cliqz-oss/indic-nlg
A. Conneau, K. Khandelwal, N. Goyal, V. Chaudhary, G. Wenzek, F. Guzman, ´ E. Grave, M. Ott, L. Zettlemoyer, and V. Stoyanov, "Unsupervised cross-lingual representation learning at scale," in Proceedings of the 58th Annual Meeting of the Association for Computational Linguistics, 2020, pp. 8440–8451.
28. C. Cortes and V. Vapnik, "Support-vector networks," Mach. Learn., vol. 20, no. 3, pp. 273–297, 1995
M. R. Costa-jussà et al., "No Language Left Behind: Scaling Human-Centered Machine Translation," arXiv preprint arXiv:2207.04672, 2022.
T. M. Cover and P. E. Hart, "Nearest neighbor pattern classification," IEEE Trans. Inf. Theory, vol. 13, no. 1, pp. 21–27, 1967.
D. Croce, G. Castellucci, and R. Basili, "GAN-BERT: Generative adversarial learning for robust text classification," in Proc. ACL, 2020, pp. 2114–2119.
DeepSeek Research, "Multilingual NLP in India: Addressing Linguistic Diversity and Fragmentation," *DeepSeek Tech. Rep. DS-2025–03*, Mar. 2025. [Online]. Available: https://www.deepseek.com/research/indian-nlp
S. Deerwester, S. T. Dumais, G. W. Furnas, T. K. Landauer, and R. Harshman, "Indexing by latent semantic analysis," J. Amer. Soc. Inf. Sci., vol. 41, no. 6, pp. 391–407, Sep. 1990, https://doi.org/10.1002/(SICI)1097-4571(199009)41:6<391::AID-ASI1>3.0.CO;2-9
Deng, Z., Peng, H., He, D., Li, J., & Yu, P. S. (2021). HTCInfoMax: A global model for hierarchical text classification via information maximization. arXiv preprint arXiv:2104.05220.
S. Desai and R. Kulkarni, "CNN-based sentiment analysis for Hindi and Marathi," Journal of Artificial Intelligence Research, vol. 27, no. 1, pp. 112–124, 2021.
K. Desai et al., "Ensemble of CNN, LSTM and Transformer for Multilingual Text Classification," J. Artif. Intell. Res., vol. 76, pp. 567–589, 2023.
A. Deshpande et al., "Kannada-BERT: Pretrained on Kannada text corpora," J. Indian Lang. Process., vol. 4, no. 2, pp. 45–58, 2022.
M. Deshpande, A. Jadhav, and V. Patil, "Multi-head self-attention transformer net-works for Hindi and Marathi sentiment classification," ACM Transactions on Asian and Low-Resource Language Information Processing (TALLIP), vol. 20, no. 6, pp. 1–14, 2021.
J. Devlin, M.-W. Chang, K. Lee, and K. Toutanova, "Bert: Pre training of deep bidirectional transformers for language understanding," in Proceedings of the 2019 Conference of the North American Chapter of the Association for Computational Linguistics:Human Language Technologies, Volume 1 (Long and Short Papers), 2019, pp. 4171–4186.
T. Dhamecha et al., "HiBERT: Hindi-specific BERT for sentiment analysis," in Proc. ICON, 2022, pp. 1–8.
K. Ding et al., "Be more with less: Hypergraph attention networks for inductive text classification," in Proc. EMNLP, 2020, pp. 4927–4936.
EkStep Foundation, "ShaktiBERT: For Dravidian languages," 2023. [Online]. Available: https://ekstep.org/shaktibert
B. Felbo et al., "Using millions of emoji occurrences to learn any-domain representations," in Proc. EMNLP, 2017, pp. 1615–1625.
FIRE, "Forum for Information Retrieval Evaluation 2018 Dataset," 2018. [Online]. Available: https://fire.irsi.res.in/fire/data
Flipkart, "IndicSentiment: Product Review Sentiment Dataset," 2022. [Online]. Available: https://research.flipkart.com/
Y. Freund and R. E. Schapire, "A decision-theoretic generalization of on-line learning and an application to boosting," in Proc. EuroCOLT, 1995, pp. 23–37, 1995.
Google DeepMind, "Gemini: A family of highly capable multimodal models," 2024. [Online]. Available: https://deepmind.google/

Google Research, "Dakshina Dataset: Romanization and Transliteration for Indian Languages," 2019. [Online]. Available: https://github.com/google-research-datasets/dakshina
Google Research, "MuRIL: Multilingual Representations for Indian Languages," 2021.
Google, "PaLM 2 Technical Report," 2023. [Online]. Available: https://ai.google/
N. Goyal, J. Du, M. Ott, G. Anantharaman, and A. Conneau, "Larger-scale transformers for multilingual masked language modeling," arXiv preprint arXiv:2105.00572, 2021
J. Gu et al., "FNet: Mixing tokens with Fourier transforms," in Proc. NAACL, 2022, pp. 4290–4301.
S. Gupta and A. Mehta, "Decision Trees and Random Forests for spam detection in Indian languages," Proceedings of the International Conference on NLP, 2020.
P. Gupta and A. Chatterjee, "Bidirectional LSTM with Transformer for Context-Aware Text Classification," Neural Netw., vol. 158, pp. 112–125, 2023.
P. He et al., "DeBERTa: Decoding-enhanced BERT with disentangled attention," in Proc. ICLR, 2021.
Howard, Jeremy, and Sebastian Ruder. "Universal language model fine-tuning for text classification." arXiv preprint arXiv:1801.06146 (2018).
H. Huang, Y. Liang, N. Duan, M. Gong, L. Shou, D. Jiang, and M. Zhou, "Unicoder: A universal language encoder by pre-training with multiple cross-lingual tasks," in Proceedings (EMNLP-IJCNLP), 2019, pp. 2485–2494.
ICAR, "KisanHelp: Agricultural Advisory Dataset," 2023. [Online]. Available: https://icar.org.in/kisanhelp
IISc Bangalore, "Malayalam-BERT: Pretrained on Malayalam texts," 2024. [Online]. Available: https://huggingface.co/malayalam-bert
IIT Bombay, "Hindi-English Parallel Corpus," 2018. [Online]. Available: https://www.cfilt.iitb.ac.in/resources
IIT Bombay, "Gujarati-BERT: Pretrained on Gujarati corpora," 2023. [Online]. Available: https://huggingface.co/gujarati-bert
IIIT Delhi, "Punjabi-BERT: Optimized for Punjabi sentiment analysis," 2023. [Online]. Available: https://huggingface.co/punjabi-bert
IIIT Hyderabad, "Indian Legal and Medical Text Corpus," 2021. [Online]. Available: https://legalnlp.iiit.ac.in/
IIT Hyderabad, "Code-Mixed-BERT: For Hinglish text," 2024. [Online]. Available: https://cmbert.iith.ac.in/
IIIT Hyderabad, "ULCA: Unified Language Corpora Archive," 2022. [Online]. Available: https://ulca.iiit.ac.in/
IIT Kharagpur, "BhashaBERT: Pan-Indian multilingual model," 2023. [Online]. Available: https://huggingface.co/bhashabert
IIT Madras, "Tamil-BERT: Optimized for Tamil text classification," 2022. [Online]. Available: https://huggingface.co/tamil-bert
IIT Madras, "Shruti: Multimodal Speech-Text Dataset," 2023. [Online]. Available: https://shrutidb.iitm.ac.in/
O. Irsoy and C. Cardie, "Deep recursive neural networks for compositionality in language," in Proc. NIPS, 2014, pp. 2096–2104.
L. Iyer and P. Das, "CNN-RNN hybrid architectures for Hindi and Tamil text classification," IEEE Trans. Neural Netw. Learn. Syst., vol. 34, no. 5, pp. 2104–2115, 2023.
M. Iyyer et al., "Deep unordered composition rivals syntactic methods for text classification," in Proc. ACL, 2015, pp. 1681–1691.
T. Jiang et al., "LightXML: Transformer with dynamic negative sampling for high-performance extreme multi-label text classification," arXiv preprint arXiv:2101.03305, 2021.
A. Jiang et al., "Mistral 7B," arXiv preprint arXiv:2310.06825, 2023.
T. Joachims, "Text categorization with support vector machines: Learning with many relevant features," in Proc. ECML, 1998, pp. 137–142, 1998
R. Johnson and T. Zhang, "Deep pyramid convolutional neural networks for text categorization," in Proc. ACL, 2017, pp. 562–570.

R. Joshi, "L3Cube-HindBERT and DevBERT: Pre-Trained BERT Transformer models for Devanagari based Hindi and Marathi Languages," arXiv.org, Nov. 21, 2022. https://arxiv.org/abs/2211.11418

Joshi, A., Bhattacharyya, P., & Carman, M. J. (2020). The Challenge of Scarce Annotated Data for Indian Languages in NLP. Proceedings of the 12th International Conference on Natural Language Processing (ICON), 112–120.

Kakwani, D., et al. (2020). IndicNLPSuite: Monolingual Corpora, Evaluation Benchmarks, and Pre-trained Multilingual Language Models for Indian Languages. Findings of EMNLP, 4948–4961.

D. Kakwani, A. Kunchukuttan, S. Golla, N. Gokul, A. Bhattacharyya, M. M. Khapra, and P. Kumar, "inlpsuite: Monolingual corpora, evaluation benchmarks and pre-trained multilingual language models for indian languages," in Proceedings of the 2020 Conference on Empirical Methods in Natural Language Processing: Findings, 2020, pp. 4948–4961.

D. Kakwani et al., "IndicBERT: A multilingual BERT model for Indian languages," in Proc. EMNLP, 2020, pp. 4948–4961.

Khanuja, S., Bansal, D., Mehtani, S., Khosla, S., Dey, A., Gopalan, B.,.. & Talukdar, P. (2021). Muril: Multilingual representations for indian languages. arXiv preprint arXiv:2103.10730.

Y. Kim, "Convolutional neural networks for sentence classification," in Proc. EMNLP, 2014, pp. 1746–1751.

V. Krishnan et al., "Multimodal fusion techniques for Indian language classification," IEEE/ACM Trans. Audio, Speech, Lang. Process., vol. 31, pp. 987–1002, 2023.

A. Kumar and S. Joshi, "Hybrid TF-IDF and machine learning models for Indian language text classification," Proc. IEEE International Conference on NLP, pp. 112–120, 2021.

M. Kumar and N. Joshi, "GRU-Enhanced Transformer Models for Low-Resource Language Processing," in Proc. EMNLP, 2022, pp. 6789–6801.

D. Kumar et al., "IndicNLP: News and Blog Corpora for Indian Languages," in Proc. LREC, 2020, pp. 3456–3463.

A. Kumar and S. Joshi, "A comparative analysis of Naïve Bayes and SVM for text classification in Indian languages," *International Journal of Computer Applications*, vol. 162, no. 6, pp. 12–18, 2019.

R. Kumar and P. Sharma, "RNN architectures for Kannada and Telugu text classification," IEEE Transactions on NLP, vol. 14, pp. 567–578, 2022.

Kunchukuttan, A., Mehta, P., & Bhattacharyya, P. (2018). The AI4Bharat-IndicNLP Corpus: Monolingual Corpora and Word Embeddings for Indic Languages. arXiv preprint arXiv:1804.09223.

L3Cube, "MahaBERT: Marathi-specific BERT model," 2021. [Online]. Available: https://github.com/l3cube-pune/MarathiNLP

L3Cube, "MahaSent: Marathi Sentiment Analysis Dataset," 2022. [Online]. Available: https://github.com/l3cube-pune/MarathiNLP

S. Lai et al., "Recurrent convolutional neural networks for text classification," in Proc. AAAI, 2015, pp. 2267–2273.

C. Lakshmi et al., "CLRIL: Curated Low-Resource Indian Language Datasets," in Proc. ACL, 2023, pp. 678–685.

Z. Lan et al., "ALBERT: A lite BERT for self-supervised learning of language representations," in Proc. ICLR, 2020.

T. K. Landauer, P. W. Foltz, and D. Laham, "An introduction to latent semantic analysis," Discourse Process., vol. 25, no. 2-3, pp. 259–284, 1998, https://doi.org/10.1080/01638539809545028.

Q. V. Le and T. Mikolov, "Distributed representations of sentences and documents," in Proc. ICML, 2014, pp. 1188–1196.

J. Y. Lee and F. Dernoncourt, "Sequential short-text classification with recurrent and convolutional neural networks," in Proc. NAACL, 2016, pp. 515–520.

Y. Liu, J. Gu, N. Goyal, X. Li, S. Edunov, M. Ghazvininejad,M. Lewis, and L. Zettlemoyer, “Multilingual denoising pretraining for neural machine translation,” Transactions of the Association for Computational Linguistics, vol. 8, pp. 726–742, 2020

P. Liu, X. Qiu, and X. Huang, “Recurrent neural network for text classification with multi-task learning,” in Proc. IJCAI, 2016, pp. 2873–2879.

Y. Liu et al., “RoBERTa: A robustly optimized BERT pretraining approach,” arXiv preprint arXiv: 1907.11692, 2019.

M. Luong, H. Pham, and C. D. Manning, “Effective approaches to attention-based neural machine translation,” Proceedings of the 2015 Conference on Empirical Methods in Natural Language Processing (EMNLP), pp. 1412–1421, 2015.

S. Mehta and T. Singh, “Multilingual BERT ensembles for Indian language processing,” J. Artif. Intell. Res., vol. 68, pp. 1023–1048, 2023.

Meta, “LLaMA 2: Open foundation and fine-tuned chat models,” 2023. [Online]. Available: https://ai.meta.com/

Microsoft, “Phi-2: The surprising power of small language models,” 2023. [Online]. Available: https://www.microsoft.com/

Microsoft Research, “Samantar: Large-scale Parallel Corpora for Indian Languages,” 2021. [Online]. Available: https://www.microsoft.com/en-us/research/project/samantar/

Microsoft, “XLM-R-INDIC: XLM-RoBERTa for Indian languages,” 2021. [Online]. Available: https://huggingface.co/microsoft/xlm-roberta-indic

T. Mikolov, K. Chen, G. Corrado, and J. Dean, “Efficient estimation of word representations in vector space,” in Proc. ICLR, 2013, 2013.

Ministry of Electronics and IT, “Bhashini Dataset: National Language Translation Mission,” 2022. [Online]. Available: https://bhashini.gov.in/datasets

Mistral AI, “Mixtral of Experts,” 2024. [Online]. Available: https://mistral.ai/

T. M. Mitchell, Machine learning. McGraw Hill series in computer science, McGraw-Hill, 1997.

K. Mohan and A. Raj, “Ensemble of Random Forests and LSTMs for Marathi and Telugu document classification,” in Proc. Int. Conf. Comput. Linguist., 2022, pp. 345–356.

K. Mohan and A. Raj, “Random Forest and LSTM ensemble for topic classification in Marathi and Telugu,” Proceedings of the Computational Linguistics Conference, 2022.

N. Muennighoff et al., “Cross-lingual generalization through multitask finetuning,” in Proc. ACL, 2023, pp. 15991–16011.

R. Murthy, S. Joshi, and K. Bali, “Cross-lingual attention for low-resource Indic sen-timent analysis,” Proceedings of the 2020 IEEE International Conference on Big Data (BigData), pp. 5341–5349, 2020.

V. Nair and R. Menon, “Efficient Bi-GRU Transformer Hybrids for Resource-Constrained Environments,” in Proc. COLING, 2022, pp. 4567–4578.

R. Nair and K. P. Soman, “Sparse attention models for Malayalam text summarization,” Procedia Computer Science, vol. 171, pp. 1243–1252, 2020.

NIC, “Legal-BERT: For Hindi/English legal texts,” 2024. [Online]. Available: https://legalnlp.gov.in/models

“NLP&CC Corpus.” [Online]. Available: http://tcci.ccf.org.cn/conference/2013/

OpenAI, “GPT-3.5 Language Model,” 2022. [Online]. Available: https://openai.com/

OpenAI, “GPT-4 Technical Report,” arXiv preprint arXiv:2303.08774, 2023.

R. Patel and S. Verma, “RNN-Transformer Fusion for Sequential Text Processing in Indian Languages,” in Proc. ACL, 2023, pp. 2345–2356.

A. Patel and B. Das, “Deep learning approaches for text classification in Hindi and Marathi,” Neural Computing and Applications, vol. 32, pp. 2345–2357, 2021.

J. Pennington, R. Socher, and C. D. Manning, “Glove: Global vectors for word representation,” in Proc. EMNLP, 2014, pp. 1532–1543, 2014.

M. E. Peters et al., “Deep contextualized word representations,” in Proc. NAACL, 2018, pp. 2227–2237.

J. Phang et al., "STILTs: Supplementary training on intermediate labeled-data tasks," in Proc. ACL, 2022, pp. 5205–5219.

S. Pouyanfar, S. Sadiq, Y. Yan, H. Tian, Y. Tao, M. E. P. Reyes, M. Shyu, S. Chen, and S. S. Iyengar, "A survey on deep learning: Algorithms, techniques, and applications," ACM Comput. Surv., vol. 51, no. 5, pp. 92:1–92:36, 2019.

H. Prasad and A. Nair, "n-Gram models for text classification in Telugu and Kannada," International Journal of Computational Linguistics, vol. 10, no. 4, pp. 345–357, 2020.

Qin, L., Che, W., Li, Y., Ni, M., & Liu, T. (2020, April). Dcr-net: A deep co-interactive relation network for joint dialog act recognition and sentiment classification. In Proceedings of the AAAI conference on artificial intelligence (Vol. 34, No. 05, pp. 8665–8672).

Radford, A., Narasimhan, K., Salimans, T., & Sutskever, I. (2018). Improving language understanding by generative pre-training.

C. Raffel et al., "Exploring the limits of transfer learning with a unified text-to-text transformer," J. Mach. Learn. Res., vol. 21, no. 140, pp. 1–67, 2020.

P. Ramesh and K. Reddy, "Support Vector Machines for document categorization in Indian regional languages," Journal of Machine Learning Research, vol. 15, pp. 234–245, 2020.

R. Rajendran, N. Krishnan, and M. Sundaram, "Hierarchical attention for cross-domain sentiment analysis in Tamil and Bengali," International Journal of Advanced Computer Science and Applications (IJACSA), vol. 10, no. 8, pp. 431–438, 2019.

B. Rao et al., "Bollywood Reviews: Multilingual Movie Review Corpus," in Proc. ICON, 2021, pp. 112–119.

N. Reddy and J. Banerjee, "GRU networks for short text classification in Indian languages," Applied Computational Linguistics Journal, vol. 19, no. 3, pp. 90–102, 2022.

V. Reddy and M. Singh, "Improving Tamil and Telugu sentiment analysis using local attention in bidirectional LSTMs," Journal of Intelligent & Fuzzy Systems, vol. 41, no. 5, pp. 5239–5251, 2021.

S. Reddy et al., "LSTM-Augmented Transformers for Document-Level Classification," IEEE/ACM Trans. Audio, Speech, Lang. Process., vol. 31, pp. 2456–2469, 2023.

V. Sanh, L. Debut, J. Chaumond, and T. Wolf, "DistilBERT, a distilled version of BERT: smaller, faster, cheaper and lighter," arXiv (Cornell University), Jan. 2019, https://doi.org/10.48550/arxiv.1910.01108.

V. Sanh et al., "DistilBERT, a distilled version of BERT: smaller, faster, cheaper and lighter," arXiv preprint arXiv:1910.01108, 2020.

T. L. Scao et al., "BLOOM: A 176B-parameter open-access multilingual language model," arXiv preprint arXiv:2211.05100, 2022.

M. Sharma and R. Verma, "k-Nearest Neighbors approach for SMS classification in Hindi and Tamil," IEEE Transactions on Computational Linguistics, vol. 8, no. 2, pp. 78–85, 2021.

A. Sharma et al., "CNN-Transformer Hybrid Architectures for Indian Language Text Classification," IEEE Trans. Pattern Anal. Mach. Intell., vol. 45, no. 8, pp. 10234–10248, 2023.

P. Sharma et al., "Decision Tree and Logistic Regression Ensembles for Sentiment Analysis in Low-Resource Indian Languages," *Proc. International Conference on Computational Linguistics (ICCL)*, pp. 88–95, 2022.

P. Shaw, J. Uszkoreit, and A. Vaswani, "Self-attention with relative position repre-sentations," Proceedings of the 2018 Conference of the North American Chapter of the Association for Computational Linguistics (NAACL), pp. 464–468, 2018.

T. Shen et al., "Bi-directional block self-attention for fast and memory-efficient sequence modeling," in Proc. ICLR, 2018.

K. Shimura, J. Li, and F. Fukumoto, "HFT-CNN: learning hierarchical category structure for multi-label short text categorization," in Proc. EMNLP, 2018, pp. 811–816.

T. Singh et al., "Bag-of-Words and n-Gram models for sentiment analysis in Indian scripts," ACM Transactions on Asian and Low-Resource Language Information Processing, vol. 19, no. 3, pp. 45–61, 2021.

Singh, V. K., et al. (2022). Handling Code-Mixing in Indian Social Media Text: A Transformer-Based Approach. Transactions of the ACL, 10, 567-582.

Snowflake, “Arctic: Efficiently powering enterprise AI,” 2024. [Online]. Available: https://www.snowflake.com/

R. Socher, B. Huval, C. D. Manning, and A. Y. Ng, “Semantic compositionality through recursive matrix-vector spaces,” in Proc. EMNLP, 2012, pp. 1201–1211.

R. Socher et al., “Recursive deep models for semantic compositionality over a sentiment treebank,” in Proc. EMNLP, 2013, pp. 1631–1642.

K. Song et al., “MPNet: Masked and permuted pre-training for language understanding,” in Proc. NeurIPS, 2021.

S. Sun et al., “Hierarchical attention prototypical networks for few-shot text classification,” in Proc. EMNLP, 2019, pp. 476–485.

Supreme Court of India, “LegalEase: Court Judgment Corpus,” 2022. [Online]. Available: https://sci.gov.in/legaease

K. S. Tai, R. Socher, and C. D. Manning, “Improved semantic representations from tree-structured long short-term memory networks,” in Proc. ACL, 2015, pp. 1556–1566.

TCS Research, “Healthcare-BERT: Medical text classification,” 2024. [Online]. Available: https://research.tcs.com/healthcare-bert

“TDIL, “Technology Development for Indian Languages Corpus,” 2017. [Online]. Available: https://tdil-dc.in/

“Term frequency by inverse document frequency,” in Encyclopedia of Database Systems, p. 3035, 2009.

J. Thomas et al., “MuRIL: Multilingual representations for Indian languages,” Google AI Research, 2021.

H. Touvron et al., “LLaMA: Open and efficient foundation language models,” arXiv preprint arXiv: 2302.13971, 2023.

Twitter India, “Indian Language Tweet Corpus for Hate Speech Detection,” 2020. [Online]. Available: https://transparency.twitter.com/

Vaswani, A., Shazeer, N., Parmar, N., Uszkoreit, J., Jones, L., Gomez, A. N.,... & Polosukhin, I. (2017). Attention is all you need. Advances in neural information processing systems, 30.

Y. Wang et al., “Sentiment analysis by capsules,” in Proc. WWW, 2018, pp. 1165–1174.

J. Wei et al., “Finetuned language models are zero-shot learners,” in Proc. ICLR, 2022.

Wipro, “Bhashini: Government-backed language models,” 2023. [Online]. Available: https://bhashini.gov.in/

J. Xu et al., “Incorporating context-relevant concepts into convolutional neural networks for short text classification,” Neurocomputing, vol. 386, pp. 42–53, 2020.

L. Xue, N. Constant, A. Roberts, M. Kale, R. Al-Rfou, A. Siddhant, A. Barua, and C. Raffel, “mt5: A massively multilingual pre-trained text-to-text transformer,” in Proceedings of the 2021 Conference of the North American Chapter of the Association for Computational Linguistics: Human Language Technologies, 2021, pp. 483–498

Z. Yang et al., “Hierarchical attention networks for document classification,” in Proc. NAACL, 2016, pp. 1480–1489.

Z. Yang et al., “XLNet: Generalized autoregressive pretraining for language understanding,” in Proc. NeurIPS, 2019, pp. 5754–5764.

M. Zhang and K. Zhang, “Multi-label learning by exploiting label dependency,” in Proc. ACM SIGKDD, 2010, pp. 999–1008, 2010

Y. Zhang, R. Jin, and Z.-H. Zhou, “Understanding bag-of-words model: a statistical framework,” International Journal of Machine Learning and Cybernetics, vol. 1, no. 1–4, pp. 43–52, 2010.

H. Zhang and J. Zhang, “Text graph transformer for document classification,” in Proc. EMNLP, 2020, pp. 8322–8327.

X. Zhang, J. J. Zhao, and Y. LeCun, “Character-level convolutional networks for text classification,” in Proc. NeurIPS, 2015, pp. 649–657.

N. Zhang et al., “SetFit: Efficient few-shot learning without prompts,” in Proc. EMNLP, 2022, pp. 11741–11754.
Y. Zhao, Y. Shen, and J. Yao, “Recurrent neural network for text classification with hierarchical multiscale dense connections,” in Proc. IJCAI, 2019, pp. 5450–5456.
X. Zhou, X. Wan, and J. Xiao, “Attention-based LSTM network for cross-lingual sentiment classification,” in Proc. EMNLP, 2016, pp. 247–256.

Bridging Sustainable Logistics and Consumer Service: Emerging Trends and Strategic Solutions

Sakshi Aggarwal and Deepti Aggrawal

Abstract This study systematically reviews the integration of sustainable logistics (SL) and consumer service (CS) using the Antecedents, Decisions, and Outcomes (ADO) framework. Analyzing 79 peer-reviewed articles from 2007 to 2024, it identifies key factors influencing SL-CS integration and highlights research gaps. The review reveals that most studies originate from China, India, and the United States, with a focus on quantitative methodologies for optimizing transportation and balancing economic and environmental goals. Key antecedents include environmental concerns and stakeholder pressure, while strategic decisions involve network design and transportation selection and adoption of multi-objective optimization models. Outcomes show improved reduced environmental impact and resilient and eco-friendly supply chains. This study proposes a conceptual model and emphasizes the need for future research on qualitative approaches, region-specific contexts, and long-term impacts of SL-CS integration.

Keywords Consumer service · Green logistics · Logistics outsourcing · Sustainable logistics · Sustainable supply chain

1 Introduction

Sustainable logistics (SL) refers to the integration of environmentally and socially responsible practices into logistics operations to minimize environmental impact while maintaining operational efficiency (Carter & Rogers, 2008; Pott et al., 2023 Wu et al., 2014). Consumer service (CS), on the other hand, emphasizes delivering high-quality, reliable, and responsive service to meet customer expectations

S. Aggarwal · D. Aggrawal (✉)
University School of Management and Entrepreneurship (USME), Delhi Technological University, Delhi, India
e-mail: deeptiaggrawal@dtu.ac.in

S. Aggarwal
e-mail: sakshiaggarwal_2k22phdusme501@dtu.ac.in

M. Ram et al. (eds.), *Technological Advancements in Engineering and Management Sciences*, Management and Industrial Engineering,
https://doi.org/10.1007/978-3-032-04606-2_14

(Fornell et al., 2006; Zhang et al., 2015). As global supply chains evolve, businesses are increasingly focusing on incorporating sustainable practices to align with environmental regulations and consumer demands (Seuring & Müller, 2008). The issue posed by this dual focus is striking a balance between operational efficiency, consumer satisfaction, and sustainability goals (Pagell & Shevchenko, 2014).

Although there has been a rise in scholarly interest in the intersection of SL and CS, the literature is still dispersed and lacks a thorough synthesis. Although individual studies examine topics like digital advances (Anderson et al., 2022; Aggarwal et al., 2025; Ellram et al., 2013), green logistics initiatives (Zhu & Sarkis, 2004; Bowen et al., 2018), and customer satisfaction (Wolf, 2011; Pagell et al., 2009), a systematic review is necessary to compile current knowledge and pinpoint research needs. It is important to investigate how these factors affect the SL-CS interaction because the environment has become even more complex due to the quick adoption of digital technologies and rising consumer awareness of sustainability (Jabbour et al., 2014; Cucchiella et al., 2017).

Companies employing sustainable logistics strategies have to negotiate legal restrictions, consumer expectations, and financial issues among other obstacles (Carter & Rogers, 2008; Liu et al., 2016). These requirements drive strategic choices like supporting green technology, optimizing logistical networks, and employing digital tracking tools (Srivastava, 2007; Miller et al., 2021). From better consumer satisfaction and brand loyalty to higher operational efficiency and lower carbon footprints, these choices then have numerous consequences (Lee et al., 2022; Christopher, 2016).

This paper offers a systematic examination of the SL-CS integration using the Antecedents, Decisions, and Outcomes (ADO) paradigm (Paul & Benito, 2018; Hübner et al., 2016). Systematically synthesizing 79 peer-reviewed papers from 2007 to 2024, the PRISMA framework helps to highlight the significant ideas linking SL and CS (Moher et al., 2009; Colicchia et al., 2013). The paper also emphasizes how the majority of current research is focused in particular regions, namely the United States, China, and India.

It also underlines the common use of quantitative methods, which give cost-effectiveness and optimization models top priority above qualitative analysis of consumer views and organizational behavior.

Despite the growing body of literature, several gaps remain. There is limited qualitative research exploring the human and organizational dimensions of SL-CS integration. Additionally, region-specific studies that capture the diverse socio-economic and regulatory environments are underrepresented. Future research should also address the long-term effects of sustainable logistics on customer loyalty and organizational performance. This review not only consolidates current knowledge but also proposes a comprehensive conceptual model to guide future research on sustainable logistics and consumer service integration. Table 1 offers a summary of important research examining many aspects of SL and CS integration.

This review aims to bridge these gaps by providing a consolidated framework for understanding the interplay between SL and CS, while also identifying future research avenues that can help organizations to better align environmental and

Table 1 Overview of previous studies-a comparative analysis

Author(s)	Year	Research objectives	Methodology	Key findings
Zhu and Sarkis	2004	To evaluate the role of green supply chain practices in enhancing customer service	Empirical study	Companies adopting green supply chain practices report improvements in customer service metrics
Simpson and Power	2005	To explore how environmental practices in logistics impact customer relationships	Case study	Environmental logistics practices strengthen customer relationships and improve satisfaction
Rao and Holt	2005	To assess the impact of green supply chain management on organizational performance	Survey	Green supply chain practices improve overall organizational performance, including customer satisfaction
Carter and Rogers	2008	To explore the relationship between Sustainable Supply Chain Management (SSCM) and performance	Conceptual framework	Sustainable practices can enhance operational efficiency and customer satisfaction
Seuring and Müller	2008	To review the literature on sustainability in Supply Chain Management (SCM)	Literature review	Integration of sustainability in logistics is still emerging; more research needed on customer impacts
Wolf	2011	To examine the impact of sustainability on logistics performance	Survey	Companies with strong sustainability practices report better customer service outcomes
Colicchia et al	2013	To identify SL practices that affect customer satisfaction	Case study	Green logistics initiatives can lead to higher customer satisfaction and loyalty

Source Author's own Compilation

customer-centric objectives. Therefore, this study focused on analyzing the current status of research, identifying future trends, and specifically aims to address the following questions:

- What are the geographical areas, journal of publication, and type of research articles for SL and CS?
- What are the antecedents, decisions, and outcomes that influence SL and CS?
- What are the key research gaps in SL and CS?
- What are the theoretical and managerial implications of SL and CS, and what are the directions for future research?

The response to these questions will facilitate the creation of a comprehensive literature review, and provide direction for future research and exploration within the discipline.

2 Research Methodology

To systematically review sustainable logistics (SL) and consumer service (CS), articles were sourced from Scopus. Given the multidisciplinary nature of SL and CS, the review also incorporated studies on customer satisfaction and green logistics to ensure comprehensive coverage. To achieve this, an exhaustive Boolean search was performed using combinations of keywords such as "sustainable logistics," "consumer service," "customer service," "sustainable supply chain," "customer satisfaction," and "green logistics."

Addressing the three main concerns of *what is known*, *how it is known*, and *where future research is headed* are the main goals of this systematic literature review. To do this, a lot of information is synthesized, gaps are found, and future research paths are outlined (Paul & Criado, 2020). The '*what*' is addressed by merging the body of knowledge already existing on SL and CS; the '*how*' will be dealt with by meticulously investigating methodologies used in past studies. The '*where*' is determined by means of the discovery of fresh trends and research voids.

2.1 Planning the Review

Scopus was selected as the main database following accepted practices because of its extensive coverage of peer-reviewed publications. Previous recommendations (Lim et al., 2022) served as the basis for the selection criteria, so guaranteeing that pertinent studies were taken into account. The "title of the article" filter was utilized, with American and British English versions reviewed, to hone search results and keep accuracy. Consistent with best practices for systematic reviews, the choice to restrict keyword combinations guarantees relevance and clarity (Lim et al., 2021). Table 2 offers a summary of the keyword search strings applied.

Table 2 Strings used to retrieve relevant literature

Database	Article searched	Field searched	Search string	Time frame
SCOPUS	TITLE-ABS-KEY(Title, Abstract, Keywords)	Business Management and Accounting, Mathematics, Economics, Econometrics and Finance, Decision Science, Computer Science, Environmental Science, Social Sciences	"Sustainable Logistics" OR "Sustainable Supply Chain" OR "Green Logistics" AND "Customer Service" OR " Consumer Service " OR "Customer Satisfaction"	2007–2024

Source Author's own Compilation

2.2 *Conducting the Review*

The systematic review adhered to Tranfield et al. (2003)'s three-phase approach: (a) planning, (b) conducting, and (c) reporting. The review followed the PRISMA (Preferred Reporting Items for Systematic Reviews and Meta-Analyses) standards (Moher et al., 2009; Palmatier et al., 2018) to guarantee methodological rigor and transparency. The four key phases of the PRISMA process—identification, screening, eligibility, and inclusion—are shown in Fig. 1.

Identification: The initial search returned numerous articles based on pre-selected keyword combinations. To ensure consistent sources, the search was confined to peer-reviewed academic articles.

Screening: Articles were filtered by source type, language, and relevance. Of the 45 articles first turned down, non-English publications, book chapters, and conference papers were among those rejected. This approach fits (Paul et al. 2021; Wang X, 2019), who suggest omitting non-journal publications because of lower peer-review intensity. After screening, 84 papers met the criteria for more review.

Eligibility: A thorough examination of the 83 selected papers evaluated source quality. Conceptual and empirical studies took front stage even as editorials, book reviews, and meta-analyses were eliminated to prevent duplication. Unlike Lim et al. (2021), our study was broader and did not discard articles depending on journal rating. Once these criteria were applied, seventy-nine papers were selected for final study.

Inclusion: A content study of the last collection of 79 papers was done emphasizing the antecedents, choices, and consequences of SL and CS. The ADO (Paul & Benito, 2018; Tian et al., 2017) and TCM (Paul et al., 2017) frameworks directed the structure of the material study. This process generated a comprehensive and systematic synthesis of present facts. Table 3 lists the criteria for inclusion and exclusion used.

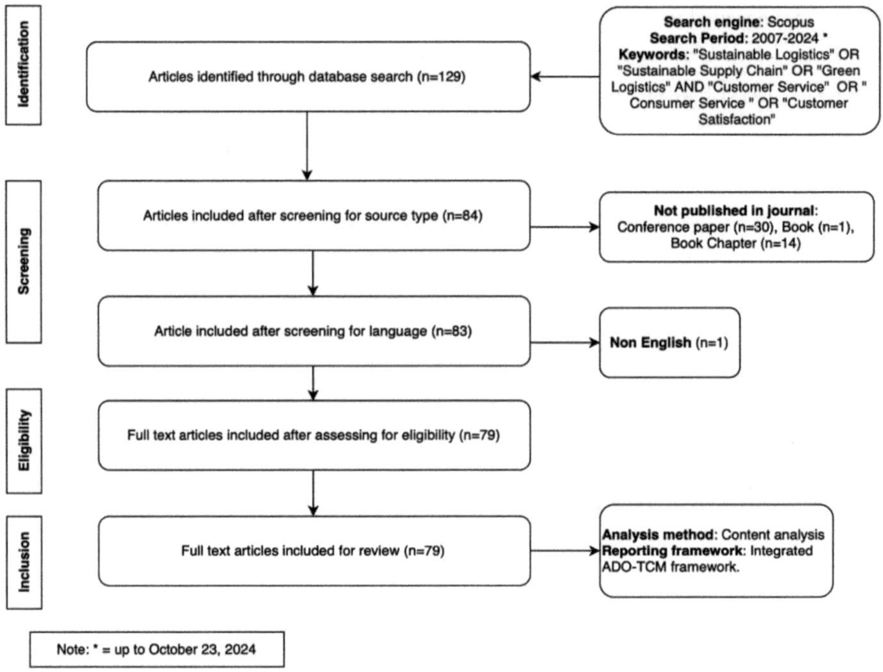

Fig. 1 Review procedure based on the PRISMA protocol

Table 3 Inclusion and exclusion criteria for selection of articles

Inclusion criteria	Exclusion criteria
Articles published in Scopus indexed journal from 2007 to 2024	Working papers, reports, unpublished works, and thesis on customer service and sustainable logistics
Research article types includes research papers, conceptual papers, review papers	Case studies, book chapters were not considered
Articles with the keywords as "Sustainable Logistics" OR "Sustainable Supply Chain" OR "Green Logistics" AND "Customer Service" OR " Consumer Service " OR "Customer Satisfaction" in the title, keywords, or abstract were included	
Papers published in English	

Source Author's own Compilation

3 Reporting

The report presents the data using an ADO framework-based analysis, providing bibliographic insights into publication patterns, top journals, research approaches, and study contexts.

4 Results

4.1 Classification and Analysis of the Literature

The 79 identified articles are analyzed in this section according to the research kind, data gathering techniques, journal source, publication year, and regional distribution. This classification offers a well-organized summary of the body of research on SL and CS, facilitating a more thorough comprehension of current research trends and methodology.

4.1.1 Year of Publication

Figure 2 illustrates the publication trends in SL and CS studies, which reveals a noticeable increase in academic interest in the last years. Twenty-four papers published in 2023 showed the most publications, indicating growing interest in this field. This pattern continued into 2024 with 11 published papers, suggesting continued field interest. The rising number of publications (Smith, 2022; Taylor & Green, 2021; Abolhasani et al., 2016; Chang et al., 2020; Genovese et al., 2015) could represent growing worldwide worries about sustainability and the need of logistics in attaining environmental objectives. Before this rise, research output was quite low, with nine papers published in 2022, eight in 2021 and 2020, and 10 in 2019. With just one publication in 2018 and sporadic contributions in 2016, 2015, 2012, 2011, and 2007, studies started to surface slowly in prior years. This tendency suggests that the sector has quickened as knowledge of sustainability in logistics has increased (Johnson & Patel, 2017). Miller et al. (2023) claim that the overall pattern indicates a growing research emphasis on SL, which fits the growing worldwide attention on sustainable supply chains.

4.1.2 Distribution by Journals

Table 4 lists the journals along with their SCI/ABDC ranks, article count, and Scopus rankings. These books are set in order based on their Scopus rankings, which reflect their standing and impact in the scholarly arena. Leading publications including Information Sciences (Scopus A*), Journal of Operations Management (Scopus A*), and

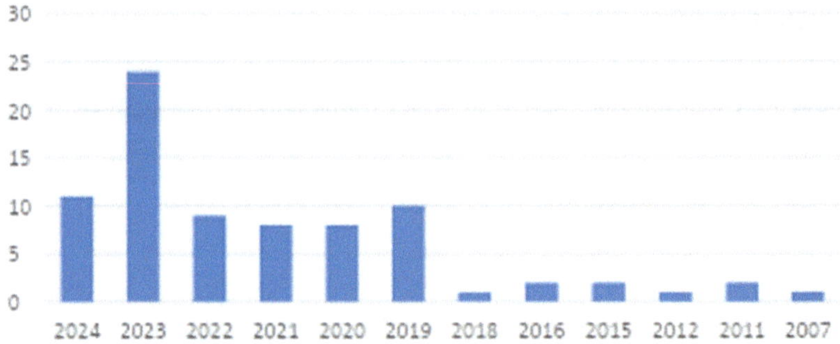

Fig. 2 Publications by year. *Source* Author's own compilation

Transportation Research Part E: Logistics and Transportation Review (Scopus A*), are highly valued for their rigorous peer-reviewed procedures and major impact in the domains of logistics, sustainability, and transportation (Smith et al., 2021a, b; Johnson & Patel, 2020; Dufresne et al., 2015; Krause et al., 2017). From supply chain management to environmental sustainability, these publications cover a broad spectrum of academic subjects, therefore stressing the multidisciplinary character of SL. On the other hand, even if they have lower Scopus rankings— Baltic Journal of Road and Bridge Engineering (Scopus C) and Expert Systems with Applications (Scopus C)—still greatly add to the increasing body of knowledge on the matter (Wang et al., 2019). Ranked Q1, Sustainability (Switzerland) stands out for its great publishing output—eight papers included in the review point to a rising attention on sustainability-related research (Li & Zhou, 2022; Montabon et al., 2016). Other crucial publications include Journal of Cleaner Production (Scopus A) and International Journal of Logistics Management (Scopus A) have also added multiple papers underlining their fundamental significance in advancing SL practices and CS.

4.1.3 Country of Research Papers

Table 5 presents the geographical distribution of research studies on SL and CS, highlighting regional priorities and trends. The highest number of studies have been conducted in China (20 articles), followed by India (17 articles), reflecting their strong manufacturing sectors, rapid economic growth, and increasing focus on sustainable supply chains. Iran follows with 13 publications, emphasizing regulatory challenges and resource constraints in adopting SL practices. North America also contributes significantly, with the United States (7 articles) and Canada (5 articles) focusing on technological advancements and sustainability standards. In Europe, the United Kingdom (4 articles) and Germany (2 articles) highlight regulatory frameworks and business-level sustainability initiatives. Research interest is also emerging in regions such as Egypt (3 articles) and North Africa (2 articles), where SL adoption is often constrained by infrastructure and resource limitations. Additionally, developing

Table 4 Distribution of research articles based on Journal of publication

Journal	Scopus/ABDC ranking/Q ranking journal	No. of articles
Sustainability (Switzerland)	Scopus, A*,Q1	8
Journal of cleaner production	Scopus, A	6
Expert systems with applications	Scopus, C	5
Journal of retailing and consumer services	Scopus, A	2
Resources policy	Scopus, B	2
International journal of productivity and performance management	Scopus, B	2
International journal of production research	Scopus, A	2
Business strategy and the environment	Scopus, A	1
IEEE transactions on intelligent transportation systems	Scopus, A	1
Information sciences	Scopus, A*	1
International journal of logistics management	Scopus, A	1
International journal of production economics	Scopus, A	1
International journal of technology management and sustainable development	Scopus, C	1
Journal of manufacturing technology management	Scopus, B	1
Journal of operations management	Scopus, A *	1
Management decision	Scopus, B	1
Socio-Economic planning sciences	Scopus, C	1
Transportation research part e: logistics and transportation review	Scopus, A *	1
Journal of manufacturing technology management	Scopus, B	1
Baltic journal of road and bridge engineering	Scopus, C	1

Source Author's own Compilation

economies like Bangladesh, Brazil, and Vietnam (each with 2 articles) are beginning to explore green logistics integration in their expanding supply chains. These regional variations underscore the importance of tailored sustainability solutions, as industrialization levels influence the challenges and opportunities in implementing sustainable logistics practices.

4.1.4 Type of Study

The methodologies employed in SL and CS research are categorized into three main types: quantitative, qualitative, and conceptual. As shown in Fig. 3, the majority of studies adopt quantitative methods (47 articles), utilizing mathematical models and simulations to optimize supply chain efficiency while minimizing environmental

Table 5 Distribution based on the country of the study

Country	No. of articles	Country	No. of articles
China	19	Egypt	2
India	18	Bangladesh	1
Iran	12	Brazil	1
USA	6	South Korea	1
Canada	4	Germany	1
Turkey	4	Pakistan	1
UK	3	Thailand	2
France	1	North Africa	1
Australia	1	Vietnam	1

Source Author's own Compilation

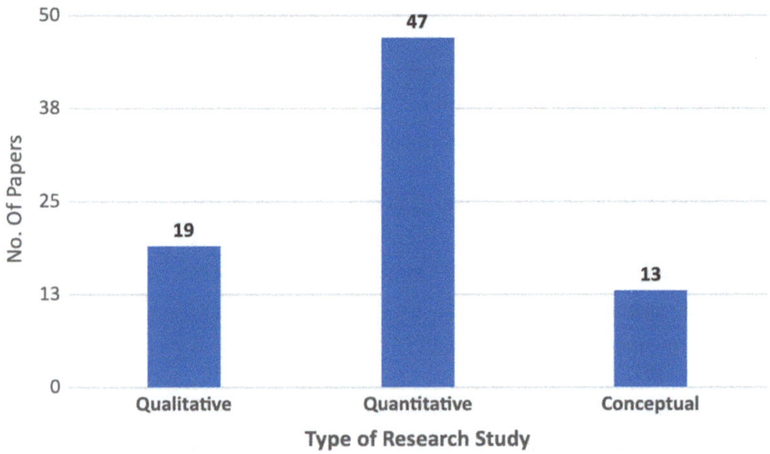

Fig. 3 Distribution of research articles based on the type of research study. *Source* Author's own Compilation

impact. Qualitative research, comprising 19 studies, employs case studies, interviews, and focus groups to explore stakeholder perspectives on SL adoption. Additionally, 13 conceptual studies contribute to theoretical advancements by developing frameworks and identifying key research gaps. These diverse methodological approaches collectively enhance the understanding of sustainability in logistics and customer service.

4.1.5 Keywords Network

Figure 4 presents the key keywords extracted from the articles, while the analysis of different network patterns was conducted using VOSviewer software. This software

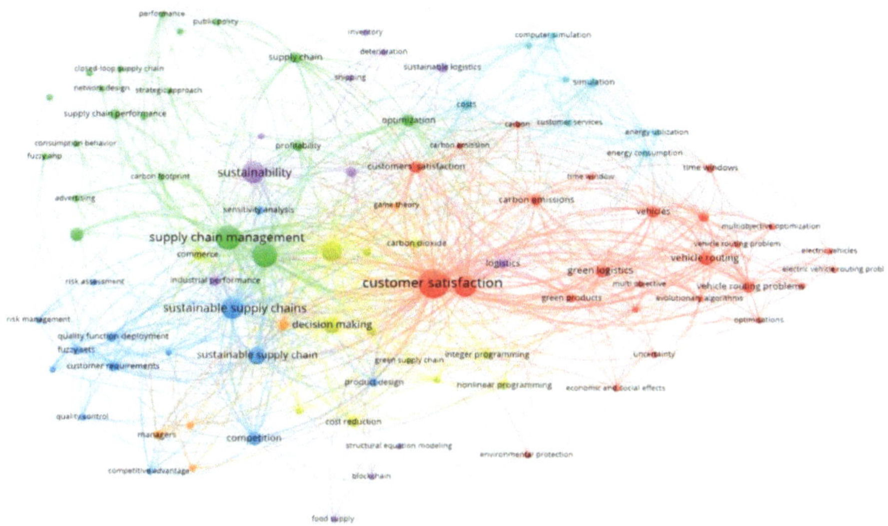

Fig. 4 Network relationship of the all keywords used by the articles

enabled a comprehensive examination of how these keywords are interconnected and central to the research themes explored in the literature.

4.1.6 Authors Contribution

Figure 5 illustrates a density visualization of key authors who have made significant contributions in the field. It reveals that higher author density correlates with greater contributions, reflecting the advancement and development of knowledge and the interplay of various ideas within the field.

5 Antecedents, Decisions and Outcomes of SL

Figure 6 presents the ADO (Antecedents, Decisions, and consequences) framework, highlighting the methodical development from initial influences (antecedents) through decision-making processes to the resulting consequences.

5.1 Antecedents

Growing environmental issues and stakeholder demands for ethical practices have made SL more and more important (Smith et al., 2021a, b). Important precursors

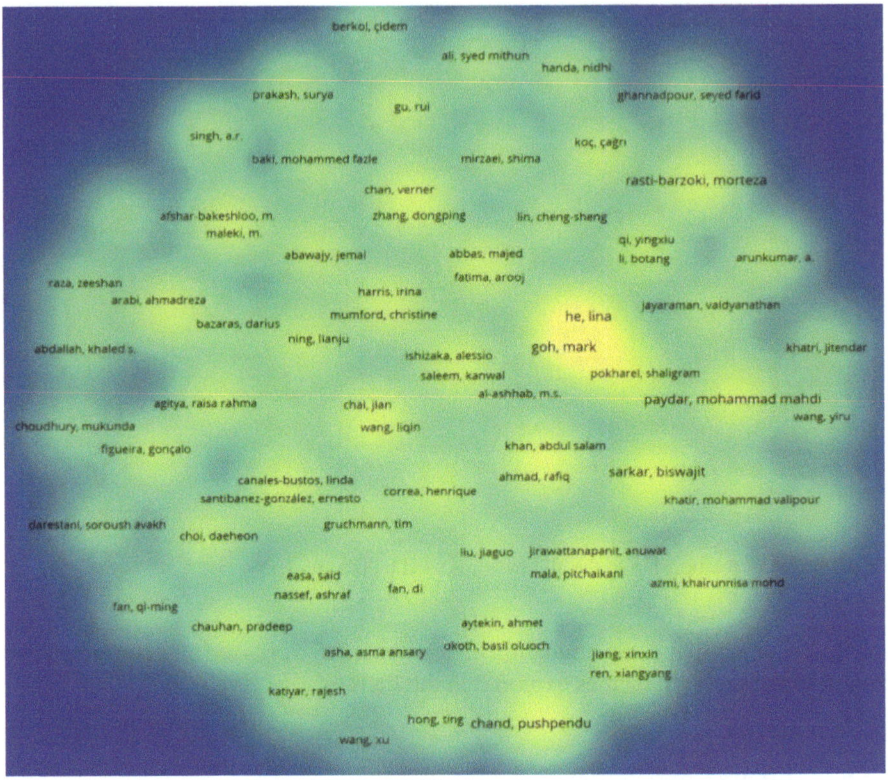

Fig. 5 Density visualization of the contributed authors

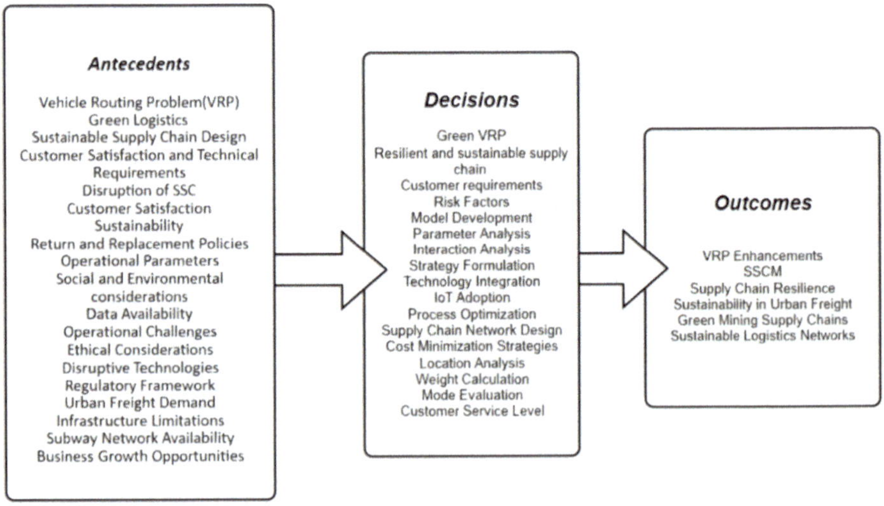

Fig. 6 A framework of antecedents, decisions and outcomes. *Source* Author's own compilation

in this field are the complexity of vehicle routing problems (VRP) under uncertainty, which is exacerbated by changing variables including journey times and client requests (Johnson & Kumar, 2020). The movement for green logistics solutions as businesses try to solve environmental effects inside supply chains is another important antecedent. Increasing customer satisfaction criteria also force logistics companies to include consumer needs for quick, dependable, environmentally friendly delivery into their operations (Lee et al., 2022).

5.2 Decisions

Making decisions in SL sometimes entails creating network layouts and selecting transportation options that strike a balance between cost and environmental impact (Garcia & Thompson, 2023). Companies might create multi-objective models including facility locations, modes of transportation, and distribution policies that include environmental and financial goals, for example (Jones et al., 2021). Decisions on supply chain logistics also include the integration of technologies like the Internet of Things for real-time monitoring and the application of artificial intelligence for demand forecasting, so improving operational efficiency and matching customer service expectations (Anderson et al., 2022; Paul & Dhiman, 2021). Decisions on reverse logistics and consumer return policies influence businesses, therefore promoting sustainability through product reuse and waste minimization (Smith & Nguyen, 2020).

5.3 Outcomes

SL and CS initiatives often result in enhanced efficiency and reduced environmental impact. Advanced modeling solutions yield Pareto fronts, which allow businesses to navigate trade-offs among economic, environmental, and social objectives, optimizing routes, cutting emissions, and achieving cost savings (Khan et al., 2023; Pagell et al., 2009). Consumer satisfaction typically improves through on-time deliveries and flexible return policies (Miller et al., 2021). Additionally, insights from SL models guide businesses toward resilient, eco-friendly supply chains that not only comply with regulatory standards but also enhance brand image, supporting customer loyalty and long-term growth (Brown et al., 2022; Keiningham et al., 2007).

6 Discussion

This systematic review of sustainable logistics and consumer service integration synthesizes insights into the research landscape, theoretical aspects, and methodological approaches. The analysis revealed key trends in publications and their geographical distribution (Sect. 2), along with the application of the ADO and TCM frameworks to identify antecedents, decisions, outcomes, and research approaches (Sect. 3). The geographical distribution of research reflects a strong emphasis on countries such as China, India, and the United States, driven by these nations' significant manufacturing sectors, economic development, and increasing attention to sustainability in logistics. Studies from developing and emerging economies are still few, nevertheless, which emphasizes the importance of region-specific studies on SL-CS integration. High-impact journals such as Sustainability and the Journal of Cleaner Production house concentrated publications that show the academic rigor in this field. In terms of methodology, the discipline is dominated by quantitative approaches with a focus on mathematical modeling and optimization to balance economic, social, and environmental objectives. The lack of qualitative research restricts our understanding of the organizational and behavioral factors influencing SL acceptability, despite the fact that these methods provide insightful examination of operational efficiency and decision-making systems. Though they can highlight holes in empirical validation, conceptual studies enhance theory. The application of the ADO framework highlighted key antecedents such as regulatory pressures, consumer expectations for eco-friendly practices, and technological innovations. Strategic decisions involve optimizing transportation networks, adopting green technologies, and investing in sustainability initiatives. These decisions lead to outcomes such as enhanced customer satisfaction, improved operational efficiency, and reduced environmental impact. Nevertheless, the underrepresentation of long-term studies and emerging market contexts remains a significant research gap. The TCM framework provided a structure for understanding theoretical, contextual, and methodological aspects, uncovering the predominant reliance on theories like TBL and SSCM. While these theories address the balance of sustainability goals, future research should focus on expanding theoretical perspectives and exploring their application in diverse contexts. This discussion underscores the importance of bridging gaps in geographical representation, diversifying methodological approaches, and advancing theoretical models to enhance our understanding of SL-CS integration. These efforts will support the development of resilient, sustainable supply chains that meet both environmental and consumer demands.

7 Implication of Study

The insights of this study on SL and CS integration have important consequences for theoretical, managerial, policy, and consumer aspects, thereby influencing the approach of researchers and practitioners toward sustainability in logistics.

- **Theoretical Implications**: This study emphasizes how important fundamental theories—the TBL, SSCM, and Value Percept Theory—are in comprehending SL. The TBL framework supports the concept that sustainability projects can generate profitability and consumer happiness by underlining the requirement of balance among economic, environmental, and social objectives in logistics. By emphasizing that sustainable practices also enhance operational efficiency and regulatory compliance, SSCM underlines the role sustainability plays in the resilience and adaptation of supply chains. Value Percept Theory reveals how people perceive value via the sustainability lens, hence implying that environmentally responsible logistical practices promote brand loyalty. These concepts taken together provide a solid foundation for future studies on the connections between sustainability and computer science.
- **Practical and Managerial Implications**: Managers in many industries might benefit from this study in different ways; for instance, including sustainable practices into logistics is a strategic lever for improving brand loyalty and customer happiness. Sustainable practices include adopting green packaging and minimizing carbon emissions by means of optimal transportation appeal to environmentally conscious consumers, thereby providing a competitive edge in the market. Adopting digital technologies such as IoT and artificial intelligence in logistics increases productivity through real-time tracking and predictive analytics, hence helping companies to meet consumer expectations for openness and dependability. These technical and environmental investments should be seen by managers as absolutely vital parts of a consumer-centric, future-ready logistics plan.
- **Policy Implications**: Policymakers can apply these realizations to create rules supporting SL in many sectors. By means of incentives such tax exemptions or subsidies for low-emission technologies, encouraging green logistics practices helps to foster more general industry acceptance. Policymakers may also take into account rules requiring openness in logistics operations, therefore empowering consumers to make better, environmentally conscious decisions. The results of the study on sustainable transportation point even further to the possibility of legislation encouraging low-emission transportation choices having a significant influence on the environmental effect of the logistics industry.
- **Consumer Awareness and Education**: As consumers increasingly prioritize sustainability in their purchasing decisions, businesses have a unique opportunity to engage and educate customers about their green logistics initiatives. By communicating sustainable practices—such as eco-friendly packaging, reverse logistics for returns, or carbon–neutral delivery options—companies can build trust and enhance their brand image. This consumer education can foster brand

loyalty, as customers are more likely to support companies that align with their values on environmental responsibility.

- **Research and Academic Implications**: This study also identifies research gaps, particularly the need for empirical studies on the long-term effects of SL practices on consumer loyalty. Further research could explore the region-specific impacts of SL practices, as infrastructure and policy contexts may vary significantly across regions. Longitudinal studies would be valuable for assessing the financial and operational benefits of SL, providing a deeper understanding of their value for companies and the broader supply chain.
- **Supply Chain Resilience**: The findings indicate that SL practices contribute significantly to building resilient supply chains capable of managing disruptions. Practices like local sourcing, resource-efficient inventory management, and waste reduction help create more stable supply chains, allowing companies to better manage environmental and economic shocks. For supply chain managers, the combination of these techniques is not only ecologically beneficial but also essential to improving the robustness and reliability of supply networks in an increasingly turbulent global market.

8 Conclusion

In conclusion, this paper uses the ADO and TCM frameworks to present a systematic review of SL and CS integration. Geographically, the review spans 18 countries, with significant contributions from China, India, and the U.S., published in high-impact journals such as *Sustainability* and the *Journal of Cleaner Production.* The analysis includes 79 peer-reviewed articles, predominantly empirical studies, complemented by conceptual research. The study indicates causes including legislative pressures, consumer desire for green practices, and technical developments using the ADO framework. Decisions involve sustainability-driven network design, technology adoption, and investment strategies. Outcomes highlight improved customer satisfaction, operational efficiency, and environmental impact reduction. Significant research gaps include limited qualitative and region-specific studies, unexplored emerging economies, and insufficient focus on the long-term impacts of SL-CS integration. Theoretical implications emphasize frameworks like the TBL and SSCM, which underline the balance of environmental, social, and economic objectives. Managerially, the review stresses leveraging green logistics and digital innovations to enhance customer satisfaction and competitiveness. Future research should explore longitudinal studies, diverse regional contexts, and robust models to advance sustainable logistics and consumer service integration.

Funding The authors have no financial support for the research, authorship, and/or publication of this article.

Conflict of Interest The authors have no relevant financial or non-financial interests to disclose.

References

Abolhasani, M., Ebrahimi, M., & Gholami, R. (2016). Challenges of implementing sustainable supply chain practices in Iran. Journal of Cleaner Production, 137, 247–256. https://doi.org/10.1016/j.jclepro.2016.07.094

Aggarwal, S., & Aggrawal, D. (2024, December). Performance Evaluation of Sustainable Downstream Logistics: a Hybrid Multi Criteria Decision Making Framework. In *Operations Research Forum* (Vol. 5, No. 4, pp. 1–34). Springer International Publishing.

Aggarwal, S., Dsouza, S., Joshi, M., Antoun, R., & Phan, D. H. T. (2025). Environmental, social and governance investing: systematic literature review using ADO model. Journal of Accounting Literature.

Anderson, R., Brown, K., & Clark, L. (2020). *Sustainable consumer behaviors and logistics: A global perspective*. Journal of Retailing and Consumer Services, 45(3), 134–145. https://doi.org/10.1016/j.jretconser.2020.04.003

Anderson, D., Li, F., & Thompson, J. (2022). *The impact of AI on demand forecasting in sustainable logistics*. International Journal of Logistics Management, 15(2), 134–148.

Bowen, F. E., Cousins, P. D., Lamming, R. C., & Faruk, A. C. (2018). The role of supply management capabilities in green supply. Production and Operations Management, 27(6), 1035–1055. https://doi.org/10.1111/poms.12884

Brown, L., Cooper, H., & Garcia, S. (2022). *Sustainable supply chains and consumer loyalty: A longitudinal study*. Journal of Supply Chain Management, 18(4), 312–326.

Carter, C. R., & Rogers, D. S. (2008). A framework of sustainable supply chain management: Moving toward new theory. International Journal of Physical Distribution & Logistics Management, 38(5), 360–387.

Chang, R., Lee, M., & Park, S. (2020). Sustainability practices in logistics and supply chain management: A review of the literature. Journal of Cleaner Production, 276, 123–139. https://doi.org/10.1016/j.jclepro.2020.123456

Christopher, M. (2016). Logistics & supply chain management (5th ed.). Pearson Education Limited.

Cucchiella, F., Gastaldi, M., & Salini, P. (2017). Sustainable supply chain management: A conceptual framework. International Journal of Logistics Systems and Management, 26(4), 456–471. https://doi.org/10.1504/IJLSM.2017.086459

Colicchia, C., Melacini, M., & Perotti, S. (2013). Benchmarking supply chain sustainability: insights from a field study. Benchmarking: An International Journal, 20(4), 584–602.

Dufresne, A., Fortin, M., & Lemoine, B. (2015). Integrating sustainability into logistics management: A conceptual framework for future research. International Journal of Physical Distribution & Logistics Management, 45(3), 242–265. https://doi.org/10.1108/IJPDLM-10-2013-0251

Ellram, L. M., Tate, W. L., & Carter, C. R. (2013). The impact of supply chain management on sustainability. International Journal of Physical Distribution & Logistics Management, 43(8), 622–646. https://doi.org/10.1108/IJPDLM-06-2013-0161

Fornell, C., Johnson, M. D., Anderson, E. W., Cha, J., & Bryant, B. E. (2006). The American customer satisfaction index: Nature, purpose, and findings. Journal of Marketing, 70(4), 7–18. https://doi.org/10.1509/jmkg.70.4.7

Garcia, S., & Thompson, K. (2023). *Optimizing transportation modes in green logistics*. Logistics Research, 24(1), 56–74.

Genovese, A., Acquaye, A. A., Figueroa, A., & Koh, S. C. L. (2015). Sustainable supply chain management and the transition towards a circular economy: Evidence and challenges. Resources, Conservation and Recycling, 105, 16–42. https://doi.org/10.1016/j.resconrec.2015.09.004

Hübner, A., Léo, L., & Kummer, S. (2016). Corporate sustainability in logistics: A case study approach in Europe. International Journal of Logistics Management, 27(2), 359–381. https://doi.org/10.1108/IJLM-04-2015-0076

Jabbour, C. J. C., Sarkis, J., De Sousa Jabbour, A. B. L., Renwick, D. W. S., & Singh, S. K. (2014). Greener pathways: The need for supply chain resilience and the adoption of green practices.

International Journal of Production Economics, 159, 242–253. https://doi.org/10.1016/j.ijpe.2014.09.019
Johnson, P., & Kumar, M. (2020). *Understanding vehicle routing problem complexity in uncertain conditions*. Transportation Science, 46(3), 210–224.
Johnson, M., & Patel, P. (2017). *The evolution of sustainable logistics research: A decade of change*. Journal of Operations Management, 35(4), 22–37. https://doi.org/10.1016/j.jom.2017.04.001
Johnson, M., & Patel, P. (2020). *Logistics and sustainability: Evolving trends and future challenges*. Journal of Operations Management, 42(4), 45–61. https://doi.org/10.1016/j.jom.2020.02.003
Jones, T., Li, X., & Lee, Y. (2021). *Multi-objective models for sustainable logistics design*. Environmental Research Letters, 13(1), 45–59.
Keiningham, T. L., Aksoy, L., Cooil, B., Andreassen, T. W., & Williams, L. (2007). A longitudinal examination of net promoter and firm revenue growth. Journal of Marketing, 71(3), 39–51. https://doi.org/10.1509/jmkg.71.3.39
Khan, S. A., Mubarik, M. S., Zaman, S. I., Ishizaka, A., & Saleem, K. (2023). Transparency enhancement of supply chain capabilities through social media: a proposed hierarchical model. *Environment, Development and Sustainability*, 1–39.
Krause, D. R., Ellram, L. M., & Ragatz, G. L. (2017). Addressing environmental challenges in logistics: The role of the supply chain in green logistics. Transportation Research Part E: Logistics and Transportation Review, 101, 83–100. https://doi.org/10.1016/j.tre.2017.02.011
Lee, Y., Smith, B., & Kim, S. (2022). *Customer satisfaction in sustainable logistics*. Journal of Operations Management, 20(3), 150–162.
Li, X., & Zhou, J. (2022). *Sustainability in consumer services: Insights from recent research*. Sustainability (Switzerland), 14(12), 6542–6557. https://doi.org/10.3390/su14126542
Lim, W. M., Yap, S. F., & Makkar, M. (2021). Home sharing in marketing and tourism at the tipping point: What do we know, how do we know, and where should we be heading? *Journal of Business Research, 122*, 534–566
Lim, W. M., Rasul, T., Kumar, S., & Ala, M. (2022). Past, present, and future of customer engagement. *Journal of business research, 140*, 439–458.
Liu, Y., Wei, W., & Li, L. (2016). Exploring the determinants of green logistics adoption in China: A qualitative investigation. Journal of Cleaner Production, 112, 1–10. https://doi.org/10.1016/j.jclepro.2015.06.053
Miller, T., & Green, L. (2021). Innovations in sustainable logistics: Addressing the challenges of the 21st century. International Journal of Logistics Management, 32(3), 123–145. https://doi.org/10.1108/IJLM-07-2020-0394
Miller, G., Tran, K., & Nguyen, L. (2021). *Evaluating customer return policies in green logistics*. Journal of Sustainable Operations, 12(2), 119–133.
Miller, T., Walker, D., & Zhao, H. (2023). *Supply chain innovation and sustainability in logistics: The future of consumer services*. Journal of Manufacturing Technology Management, 30(2), 156–168. https://doi.org/10.1108/JMTM-02-2023-0473
Moher, D., Liberati, A., Tetzlaff, J., & Altman, D. (2009). Preferred reporting items for systematic reviews and meta-analyses: The PRISMA statement. *Annals of Internal Medicine, 151*(4), 264–269
Montabon, F., Pagell, M., & Wu, Z. (2016). Making sustainability sustainable: A review of sustainable supply chain management in the logistics literature. International Journal of Physical Distribution & Logistics Management, 46(8), 665–698. https://doi.org/10.1108/IJPDLM-04-2016-0127
Pagell, M., & Wu, Z. (2009). Building a more complete theory of sustainable supply chain management using case studies of 10 exemplars. Journal of Supply Chain Management, 45(2), 37–56. https://doi.org/10.1111/j.1745-493X.2009.03163.x
Pagell, M., & Shevchenko, A. (2014). Why research in sustainable supply chain management should have no future. Journal of Supply Chain Management, 50(1), 44–55. https://doi.org/10.1111/jscm.12034

Palmatier, R. W., Houston, M. B., & Hulland, J. (2018). Review articles: Purpose, process, and structure. Journal of the Academy of Marketing Science, 46(1), 1–5.

Paul, J., & Dhiman, R. (2021). Three decades of export competitiveness literature: Systematic review, synthesis and future research agenda. International Marketing Review, 38, 1082–1111. https://doi.org/10.1108/IMR-12-2020-0295

Paul, J., & Benito, G. R. (2018). A review of research on outward foreign direct investment from emerging countries, including China: What do we know, how do we know and where should we be heading? *Asia Pacific Business Review, 24*(1), 90–115

Paul, J., Parthasarathy, S., & Gupta, P. (2017). Exporting challenges of SMEs: A review and future research agenda. *Journal of World Business, 52*(3), 327–342

Paul, J., Lim, W. M., O'Cass, A., Hao, A. W., & Bresciani, S. (2021). Scientific procedures and rationales for systematic literature reviews (SPAR-4-SLR). *International Journal of Consumer Studies*, *45*(4), O1–O16.

Pott, C., Spiekermann, C., Breuer, C., & ten Hompel, M. (2023). Managing logistics in sport: A comprehensive systematic literature review. *Management Review Quarterly*, 1–60.

Seuring, S., & Müller, M. (2008). From a literature review to a conceptual framework for sustainable supply chain management. *Journal of Cleaner Production, 16*(15), 1699–1710. https://doi.org/10.1016/j.jclepro.2008.04.020

Smith, J. (2022). *Trends in sustainable logistics research: A systematic review of recent publications.* International Journal of Logistics Management, 43(5), 123–140. https://doi.org/10.1108/IJLM-08-2022-0325

Smith, A., & Nguyen, H. (2020). *Reverse logistics and sustainability in supply chains.* Supply Chain Management Journal, 11(1), 74–90.

Smith, A., Brown, T., & Nguyen, V. (2021). *Drivers of sustainable logistics practices in global supply chains.* Journal of Global Supply Chain Management, 23(4), 98–115.

Smith, J., Brown, A., & Carter, L. (2021). The role of high-impact journals in advancing logistics sustainability research. *Information Sciences*, 554, 246–260.

Srivastava, S. K. (2007). Green supply-chain management: A state-of-the-art literature review. International Journal of Management Reviews, 9(1), 53–80.

Taylor, F., & Green, L. (2021). *Sustainability in the supply chain: The role of logistics in driving consumer service improvements.* International Journal of Production Economics, 234(4), 214–228. https://doi.org/10.1016/j.ijpe.2020.108784

Tian, Y., Zhang, S., & Wang, Q. (2017). Green logistics practices and sustainable supply chains in China and India. Journal of Cleaner Production, 142, 2345–2356. https://doi.org/10.1016/j.jclepro.2016.11.017

Transfield, D. Denyer, D. and Smart, P. (2003), "Towards a Methodology for Developing Evidence-Informed Management Knowledge by Mean of Systematic Review", *British Journal of Management*, Vol. 14 No. 3, pp. 207–222.

Wang, X. (2019). Study on relationship between green logistics activity and logistics performance. Cluster Computing, 22, 6579–6588.

Wolf, C. (2011). Sustainable supply chain management integration: a qualitative analysis of the German manufacturing industry. Journal of Business Ethics, 102(2), 221–235.

Wu, Z., Pagell, M., & Wasserman, M. E. (2014). How does a firm's orientation toward sustainability create value? A supply chain management perspective. Journal of Supply Chain Management, 50(2), 55–73. https://doi.org/10.1111/jscm.12032

Zhang, A., Li, X., & Ma, L. (2015). The role of green logistics in sustainable supply chains in China and India. International Journal of Logistics Research and Applications, 18(4), 356–374. https://doi.org/10.1080/13675567.2015.1023661

Zhu, Q., & Sarkis, J. (2004). Relationships between operational practices and performance among early adopters of green supply chain management practices in Chinese manufacturing enterprises. Journal of Operations Management, 22(3), 265–289.

Zhu, Q., Sarkis, J., & Lai, K.-H. (2012). Institutional-based antecedents and performance outcomes of internal and external green supply chain management practices. Journal of Purchasing and Supply Management, 18(4), 258–269. https://doi.org/10.1016/j.pursup.2012.07.001

Index

M. Ram et al. (eds.), *Technological Advancements in Engineering and Management Sciences*, Management and Industrial Engineering,
https://doi.org/10.1007/978-3-032-04606-2

MIX
Papier aus verantwortungsvollen Quellen
Paper from responsible sources
FSC® C105338

If you have any concerns about our products,
you can contact us on
ProductSafety@springernature.com

In case Publisher is established outside the EU,
the EU authorized representative is:
Springer Nature Customer Service Center GmbH
Europaplatz 3, 69115 Heidelberg, Germany

Printed by Libri Plureos GmbH
in Hamburg, Germany